Introduction to Practical Fluid Flow

This book is dedicated to my
wife Ellen

Introduction to Practical Fluid Flow

R.P. King

University of Utah

BUTTERWORTH
HEINEMANN

OXFORD AMSTERDAM BOSTON LONDON NEW YORK PARIS
SAN DIEGO SAN FRANCISCO SINGAPORE SYDNEY TOKYO

Butterworth-Heinemann
An imprint of Elsevier Science
Linacre House, Jordan Hill, Oxford OX2 8DP
200 Wheeler Road, Burlington, MA 01803

First published 2002

British Library Cataloguing in Publication Data
King, R.P.
Introduction to practical fluid flow
1 Fluid dynamics
I Title
620.1′064

Library of Congress Cataloguing in Publication Data
King, R.P.
Introduction to practical fluid flow / R.P. King.
p. cm.
Includes bibliographical references and index.
ISBN 0 7506 4885 6
1 Fluid dynamics I Title
TA357 .K575 2002
620.1′064–dc21 2002029940

ISBN 0 7506 4885 6

For information on all Butterworth-Heinemann publications visit our website at www.bh.com

Typeset by Integra Software Services Pvt. Ltd, Pondicherry 605 005, India
www.integra-india.com

Printed and bound in Great Britain by Biddles Ltd, *www.biddles.co.uk*

Contents

Preface

This book deals with the transportation and handling of incompressible fluids. This topic is important to most process engineers, because large quantities of material are transported in the process engineering industries. The emphasis of this book is on suspensions of particulate solids although the basic principles of simple Newtonian fluid flow form the basis of the development of models for the transportation of such material. Both settling slurries and dense suspensions are considered. The latter invariably exhibit non-Newtonian behavior. Transportation of slurries and other non-Newtonian fluids is generally treated inadequately or perfunctorily in most of the texts dealing with fluid transportation. This is a disservice to modern students in chemical, metallurgical, civil, and mining engineering, where problems relating to the flow of slurries and other non-Newtonian fluids are commonly encountered. Although the topics of non-Newtonian fluid flow and slurry transportation are comprehensively covered in specialized texts, this book attempts to consolidate these topics into a consistent treatment that follows naturally from the conventional treatment of the transportation of incompressible Newtonian fluids in pipelines. In order to keep the book to a reasonable length, solid–liquid systems that are of interest in the mineral processing industries are emphasized at the expense of the many other fluid types that are encountered in the process industries in general. This reflects the particular interests of the author. However, the student should have no difficulty in adapting the methods that are described here to other application areas. The level is kept to that of undergraduate courses in the various process engineering disciplines, and this book could form the basis of a one-semester course for students who have not necessarily had exposure to formal fluid mechanics. This book could also usefully be adopted for students who have or will take a course in fluid mechanics and who need to explore the typical situations that they will meet as practising process engineers. The level of mathematical analysis is consistent with that usually found in modern undergraduate engineering curricula and is consistent with the need to describe the subject matter at the level that is used in modern engineering analysis. Modeling methods that are based on partial differential equations are used in Chapter 6 because they are essential for the proper description of industrial sedimentation and thickening processes where the solid concentration frequently varies spatially and with time.

An important novel feature of this book is the unified treatment of the friction factor information that is used to calculate the flow of all types of fluid in round pipes. For each of the fluid types that are studied, the friction factor is presented graphically in terms of the appropriate Reynolds number, the dimensionless pipe diameter, the dimensionless flowrate and the dimensionless flow velocity. Each of these graphical representations leads to the most convenient computational method for specific problems depending on what information is specified and which variables must be computed. The same problem-solving methods are used irrespective of the type of fluid be it a simple Newtonian or a rheologically complex fluid such as those whose behavior is described by the Sisko model. This uniformity should assist students considerably in learning the basic principles and applying them across a wide range of application areas.

The presentation of material is somewhat different to that found in most textbooks in this field in that it is acknowledged that modern students of engineering are computer literate. These students are accustomed to using spreadsheets and other well-organized computational aids to tackle technical problems. They do not rely only on calculators and almost never plot graphs using pencil and paper. Few students submit handwritten reports. Consequently, computer-oriented methods are emphasized throughout, and, where appropriate, time-consuming or tedious computational processes are preprogrammed and made available in the computational toolbox that accompanies this text. This toolbox has been designed with care to ensure that it does not provide point-and-click solutions to problems. Rather the student is encouraged to formulate a solution method for every specific problem, but the tools in the toolbox make it feasible to tackle realistic problems that would be simply too time consuming using manual computational methods or if the student were required to generate the appropriate computer code. In any case, students of process engineering are becoming less fluent in the traditional computational languages Fortran, C, Basic, and Pascal that almost all could use with some degree of proficiency during the last three decades of the twentieth century. Now, engineering students are far more likely to be fluent in computer languages such as Java and HTML and are more likely to be able to create a website on the Internet than to be able to quickly and correctly integrate a couple of differential equations numerically. Nevertheless, they are well-attuned to using solution methods that are preprogrammed and ready to be used. Students and instructors are encouraged to install the toolbox and to explore its constituent tools before tackling any material in this book. No specific programming skills are required of the student or the instructor. The use of this modern problem-solving methodology makes it possible to extend the treatment from a purely superficial level to a more in-depth treatment and so equip the student to tackle, and successfully solve, realistic engineering problems.

The quantitative models that are described in this text will surely change and evolve over the years ahead as a result of continuing research and investigational effort. However, the basic approach should be sufficiently general to accommodate these developments. Because the computational toolbox has an open-ended design, new models can be inserted with ease at any time and it is intended that the toolbox should continue to expand well into the future.

This book can be used as a reading text to support Internet-based course delivery. This method has been used with success at the University of Utah, where such a course, supported by a fully equipped virtual laboratory, is now available. At the time of writing this course can be previewed at http://webct.tacc.utah.edu.

Professor R.P. Chabbra and Professor Raj Rajamani made several useful suggestions for improving the first draft of this book. These are gratefully acknowledged.

R.P. King
Salt Lake City

1
Introduction

1.1 Fluid flow in process engineering

Process engineering deals with the processing of large quantities of material. In order to process materials, they must be transported to the processing plant, and they must be transported from one unit operation to another within the processing environment. Materials are usually transported in a fluid phase, because this is generally much easier and more cost-effective than transportation as a solid. Liquids can be easily moved through pipelines or open channels, and the energy that is required can be conveniently delivered to the fluid using a pump. Gases too can be transported economically by pipeline, but this book deals exclusively with the transportation of incompressible fluids. These include pure liquids, both Newtonian and non-Newtonian, and suspensions of solid particles in liquids that form slurries or pastes. Non-Newtonian fluids and suspensions are commonly encountered by chemical, metallurgical, mining, and civil engineers.

This book does not start from the usual definition of the fluid as a continuum from which the application of differential mass and momentum balances leads to the equation of continuity and the Navier–Stokes equations. The approach taken is macroscopic with an emphasis of solving problems of practical engineering significance. The accompanying computational toolbox provides the tools that are necessary to solve these problems with convenience, and the reader is expected to become familiar with the toolbox and its contents.

1.2 Dimensions, units, and physical quantities

A variety of physical quantities of both the fluids and the equipment will be used throughout this book. These quantities must be described quantitatively, for which sets of dimensions and units are required. For example, the density of a fluid is an important quantity that will influence the behavior of the fluid in most situations. The dimensions of density are mass per unit volume M/L^3. To give the density a numerical value, a set of units must be selected for all the dimensions that are to be used. In this book all units will be specified in the SI (Système International) system. There is a good reason for this: the SI is the only practical set of units that is coherent. This means that no conversion factors are ever required when solving problems. This is in stark contrast to all other systems of units,

Table 1.1 Fundamental dimensions in the SI and their units

Quantity	*Dimension*	*SI unit*	*Symbol*
Length	L	meter	m
Mass	M	kilogram	kg
Time	T	second	s
Electric current		ampere	A
Temperature	K	kelvin	K
Quantity of a substance	M	gram-mole	mol
Luminous intensity		candela	cd
Plane angle		radian	rad
Solid angle		steradian	sr

including the metric system, which require difficult-to-remember conversion factors in almost every problem except perhaps only the most elementary and trivial. These older incoherent systems of units are now regarded as being obsolete for the purposes of scientific and technical calculations.

The SI is based on a set of fundamental dimensions and units as shown in Table 1.1. The precise size of each of the fundamental dimensions is defined by reference to a unique physical entity. Because the size of the fundamental dimensions that are used in the SI do not always conveniently match those of the physical quantities that are encountered in practical problems, a set of prefixes is defined which specify powers of 10 which multiply the fundamental units as required for convenient specifications of the numerical quantities. These are given in Table 1.3.

Clearly, the fundamental dimensions are not sufficient to describe all the physical properties that are of interest, and a set of derived units that will be of interest in this book is given in Table 1.2.

For example, the unit of density in the SI system is kg/m^3.

The use of upper case letters in the unit abbreviations is restricted to those units that are named for people. In Table 1.2 these are the newton (N), hertz (Hz), pascal (Pa), joule (J), watt (W) and kelvin (K).

Some units that are outside the SI but which may be used with the SI are given in Table 1.4. These outside units are not coherent with the SI and should never be used in calculations. Convert any quantity in these units to the SI unit before calculations begin.

The coherence of the SI system is demonstrated using the following simple example. The energy that is required to transport a fluid from one location to another can be calculated using the following equation, which is derived in Chapter 2.

$$\begin{aligned}\text{Energy required} = {} & \text{Change in potential energy} + \text{Change in kinetic energy} \\ & + \text{specific volume of fluid} \times \text{Change in pressure} \\ & + \text{Energy dissipated by friction.}\end{aligned}$$

Table 1.2 Some derived units in the SI

Quantity	*Dimension*	*SI unit*	*Name*
Area	L^2	m^2	
Volume	L^3	m^3	
Velocity	L/T	m/s	
Acceleration	L/T^2	m/s^2	
Angular velocity	T^{-1}	rad/s	
Force	ML/T^2	N	newton
Density	M/L^3	kg/m^3	
Frequency	T^{-1}	Hz	hertz
Pressure	M/LT^2	$Pa = N/m^2$	pascal
Specific energy	L^2/T^2	J/kg	
Stress	M/LT^2	N/m^2	
Surface tension	M/T^2	N/m	
Work	ML^2/T^2	J = Nm	joule
Energy	ML^2/T^2	J = Nm	joule
Torque	ML^2/T^2	Nm	
Power	ML^2/T^3	Nm/s = J/s = W	watt
Entropy	ML^2/T^2 K	J/K	
Viscosity	M/LT	kg/m s = Pa s	
Mass flow	M/T	kg/s	
Volume flow	M^3/T	m^3/s	

Such an energy balance is usually established for unit mass of fluid that flows. The energy required will now be calculated using obsolete units and SI units to demonstrate the advantages that are gained through the coherence of the latter system.

Table 1.3 SI prefixes

Multiplying factor	*Prefix*	*Symbol*
10^{12}	tera	T
10^9	giga	G
10^6	mega	M
10^3	kilo	k
10^{-2}	centi	c
10^{-3}	milli	m
10^{-6}	micro	μ
10^{-9}	nano	n
10^{-12}	pico	p

Table 1.4 Some units outside the SI that are accepted for use with the SI

Name	*Symbol*	*Value in SI units*
minute (time)	min	1 min = 60 s
hour	h	1 h = 60 min = 3600 s
day	d	1 d = 24 h = 86400 s
degree (angle)	°	1° = (π/180) rad
liter	L	1 L = 10^{-3} m^3
metric ton	t or tonne	1 t = 1000 kg
bar	bar	1 bar = 0.1 Mpa = 100 kPa = 10^5 Pa

Table 1.5 Data for illustrative example

Data	*Obsolete units*	*SI units*
Initial elevation	3 ft above datum	0.9144 m
Final elevation	25 ft above datum	7.620 m
Initial velocity	2 ft/sec	0.6096 m/s
Final velocity	5 ft/sec	1.5240 m/s
Initial pressure	65 psig	4.482 × 10^5 Pa
Final pressure	0 psig	0 Pa
Energy dissipated by friction	0.253 Btu/lb_m	5.88.48 J/kg
Density of fluid	62.4 lb_m/ft^3	999.52 kg/m^3
Gravitational acceleration	32.2 ft/sec^2	9.8081 m/s^2
Atmospheric pressure	740 mm mercury	98.664 kPa

The data for this example is set out in Table 1.5. The standard method for setting out this calculation in the old system of units, as taught in many high schools and universities in the United states, is as follows:

$$\text{Energy required} = g(z_{\text{final}} - z_{\text{initial}}) + \frac{1}{2}(V_{\text{final}}^2 - V_{\text{initial}}^2) + \frac{(P_{\text{final}} - P_{\text{initial}})}{\rho} + F$$

$$= \frac{32.2\ \text{ft}}{s^2}\left|\frac{(25-3)\ \text{ft}}{}\right|\frac{1\ \text{lb}_f}{32.174\ \text{lb}_m\ \text{ft/s}^2}$$

$$+ \frac{0.5(5^2-2^2)\ \text{ft}^2/\text{s}^2}{}\left|\frac{1\ \text{lb}_f}{32.174\ \text{lb}_m\ \text{ft/s}^2}\right.$$

$$+ \frac{}{62.4\ \text{lb}_m/\text{ft}^3}\left|\frac{(0-65)\ \text{lb}_f/\text{inch}^2}{}\right|\frac{12^2\ \text{inch}^2/\text{ft}^2}{}\bigg|$$

$$+ \frac{15.3\ \text{Btu/lb}_m}{}\left|\frac{1\ \text{ft-lb}_f}{1.284\times 10^{-3}\ \text{Btu}}\right.$$

$$= 22.02\ \text{ft-lb}_f/\text{lb}_m + 0.326\ \text{ft-lb}_f/\text{lb}_m - 150.00\ \text{ft-lb}_f/\text{lb}_m + 197.04\ \text{ft-lb}_f/\text{lb}_m$$

$$= \frac{69.38\ \text{ft-lb}_f/\text{lb}_m}{}\left|\frac{1.284\times 10^{-3}\ \text{Btu}}{1\ \text{ft-lb}_f}\right.$$

$$= 0.0891\ \text{Btu/lb}_m$$

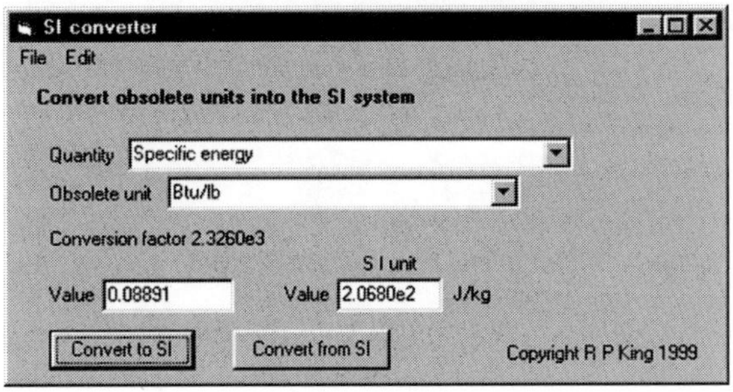

Figure 1.1 SI unit converter in the FLUIDS toolbox

The above method is error prone, time consuming, and totally unnecessary if the SI is used.

Using SI units, the calculation is set out in the following simple and intuitive form by substituting the numerical values directly for the symbols in the formulas.

$$\begin{aligned}\text{Energy required} &= 9.8081(7.620 - 0.9144) + 0.5(1.524^2 - 0.096^2) \\ &\quad + \frac{(0 - 4.482 \times 10^5)}{999.52} + 588.48 \\ &= 206.8\,\text{J/kg}\end{aligned}$$

Because SI units are used throughout, the units of the separate terms are automatically consistent because of the coherence of the SI. Each term represents energy per unit mass so the answer is automatically in the SI unit for this quantity, namely J/kg. No conversion factors are required.

This simple example suggests an effective strategy for dealing with calculations when the original data is specified in obsolete units. First convert all the primary data into SI units. Perform all calculations in SI units which will never require any conversion factors. The final answer is always in the appropriate SI unit. If required, the final answer can be reported in any other system of units by doing a single conversion out of SI to whatever unit is required. The conversion of units into and out of the SI system is facilitated by the SI conversion feature that is included in the FLUIDS toolbox on the CD-ROM that is included with this book. This converter is illustrated in Figure 1.1.

1.3 Properties of fluids

Some elementary physical properties of fluids are discussed in this section.

1.3.1 Density and specific gravity

The density of a fluid is the mass of a unit volume of the fluid. For example, water has a density of 998 kg/m^3 at 20 °C. The density is usually represented by the symbol ρ.

The specific gravity of any substance is the ratio of the density of the substance to the density of water. Specific gravity is usually represented by the symbol s.

$$s = \frac{\rho}{\rho_{\text{water}}} \tag{1.1}$$

1.3.2 Viscosity

Viscosity can be thought of as the internal stickiness of a fluid. When a fluid flows, it deforms as one layer of fluid flows over another, and the rate of deformation is governed by internal shearing stresses that are set up within the fluid. The relationship between the shear stress and the rate of deformation for many fluids is governed by the simple linear relationship

$$\tau = \mu \frac{du}{dy} \tag{1.2}$$

Where τ represents shear stress, μ represents the viscosity, u is the velocity in the fluid, and y is a spatial coordinate. The quantity du/dy is a velocity gradient and can be interpreted as the rate at which strain or extension develops in the fluid. Fluids that are described by equation 1.2 are known as Newtonian fluids. Other types of behavior are possible, and fluids that deviate from equation 1.2 are called non-Newtonian. Such fluids are discussed in detail in Chapter 5.

The viscosity of a fluid is quite sensitive to the temperature, and liquids show a strong decrease in viscosity as temperature increases.

1.3.3 Vapor pressure

All liquids show a greater or lesser tendency to vaporize and, if allowed to come to equilibrium with its surroundings, a liquid will establish an equilibrium across the liquid–vapor interface. The pressure exerted by the molecules of the fluid in the vapor phase is specific to each liquid, and the equilibrium pressure is called the vapor pressure of the liquid. The vapor pressure is a function of the temperature.

If the vapor pressure of a liquid exceeds the prevailing total pressure, the liquid vaporizes rapidly by boiling. This phenomenon is commonly encountered with water, which has a vapor pressure of 101.3 kPa at 100 °C. This is the pressure that is exerted by the atmosphere of the earth at sea level. Water boils briskly when in an open container at this temperature.

1.4 Fluid statics

When a fluid is stationary with respect to its container, there is no relative motion between any neighboring elements in the fluid. However, conditions in the fluid are not uniform, since it responds to the gravitational force field and the pressure in the fluid increases in the direction of the force field (see Figure 1.2). In the absence of any other forces, such as centrifugal force for

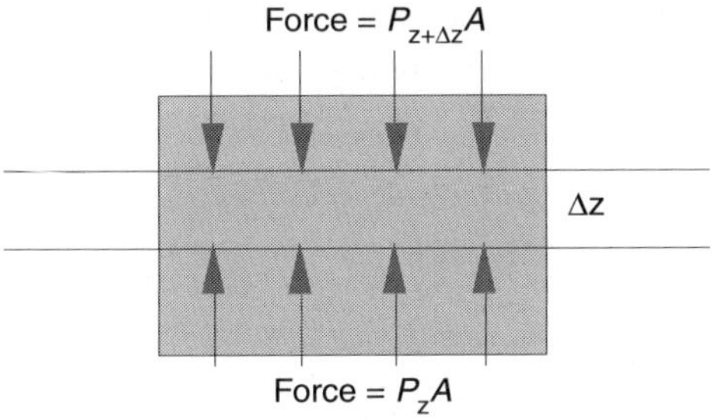

Figure 1.2 Vertical variation of pressure in a static fluid

example, the pressure is uniform on any horizontal cross-section. The pressure variation in the vertical direction is governed by a differential equation, which can be derived simply by considering a force balance over a thin imaginary horizontal slice in the body of the fluid.

Since the fluid is in equilibrium, the weight of the slice of liquid must be balanced by the net pressure force,

$$P_{z+\Delta z}A + \Delta z A \rho_f g = P_z A \tag{1.3}$$

$$\frac{P_{z+\Delta z} - P_z}{\Delta z} = -\rho_f g \tag{1.4}$$

where ρ_f is the density of the fluid.

In the limit as $\Delta z \to 0$, this generates the differential equation

$$\frac{dP}{dz} = -\rho_f g \tag{1.5}$$

Over modest depths in the gravitational fields of the earth, ρ_f and g are constant, and this equation can be integrated to give

$$P = -\rho_f g z + \text{constant} \tag{1.6}$$

If h is the distance below the free surface of the fluid, the pressure is given by

$$P = g\rho_f h + \text{atmospheric pressure} \tag{1.7}$$

The unit of pressure in the SI is the pascal (Pa), which is identical to N/m^2.

1.5 Practice problems

1. Water of density $\rho = 62.4\,\text{lbs/ft}^3$ flows at velocity $V = 4.2\,\text{ft/s}$ through a pipe of diameter $D = 6$ inches. The viscosity of the water is 1 centipoise. Calculate the value of the Reynolds number

$$Re = \frac{DV\rho}{\mu}$$

2. Show that the Reynolds number has no dimensions.

3. Calculate the pressure 10 ft below the surface of a swimming pool situated at sea level. Atmospheric pressure is 1005 millibar and the density of water is 998 kg/m^3.

1.6 Symbols used in this chapter

F	Energy dissipation by friction J/kg
g	Gravitational field
K	Kelvin
L	Length
M	Mass
P	Pressure
s	Specific gravity
T	Time
u	Velocity in fluid
V	Velocity
z	Vertical height
μ	Viscosity
ρ	Density
τ	Shear stress

Bibliography

The United States standard for SI units is defined by the National Institute for Standards and Technology (NIST). At the time of writing a summary of this standard is available at http://physics.nist.gov/cuu/Units/

Factors for converting obsolete units to SI units are given by Lees (1968).

Reference

Lees, F.P. (1968). An SI unit conversion table for chemical engineers. *The Chemical Engineer*, October 1968, pp. CE341–CE344.

2

Flow of fluids in piping systems

2.1 Pressure drop in pipes and channels

When a fluid flows through a pipe, it transfers momentum to the pipe wall. In other words the pipe wall experiences a force in the direction in which the fluid is moving. This force is best thought of as a frictional drag on the inside of the pipe wall. This concept is illustrated in Figure 2.1. The frictional drag acts along the inner surface of the pipe wall and appears as a shearing stress. The fluid against the inner pipe surface experiences this shearing stress which resists the motion of the fluid. The force on the inside wall of the pipe is related to the shearing stress through the relationship

$$\text{Force} = \pi D L \tau_w \text{ N} \tag{2.1}$$

where D is the pipe diameter and L is the length of the pipe. The shearing stress on the wall, τ_w, is determined primarily by the velocity of the fluid in the pipe but it is also a function of the properties of the fluid (density and viscosity). Obviously it is reasonable to expect that a more viscous fluid will exert a greater shearing stress than a mobile fluid such as water. The shear stress is related to the fluid velocity through an empirical equation

$$\tau_w = \left(\frac{1}{2}\rho_f \bar{V}^2\right) f \text{ N/m}^2 \tag{2.2}$$

where $\bar{V}$ is the average velocity of the fluid in the pipe, ρ_f is the fluid density and f is a variable called the friction factor. The friction factor is a function of the properties of the fluid but its value depends mostly on the state of turbulence in the fluid. Interestingly, the friction factor decreases as the fluid becomes more intensely turbulent dropping from a value around 0.012 for

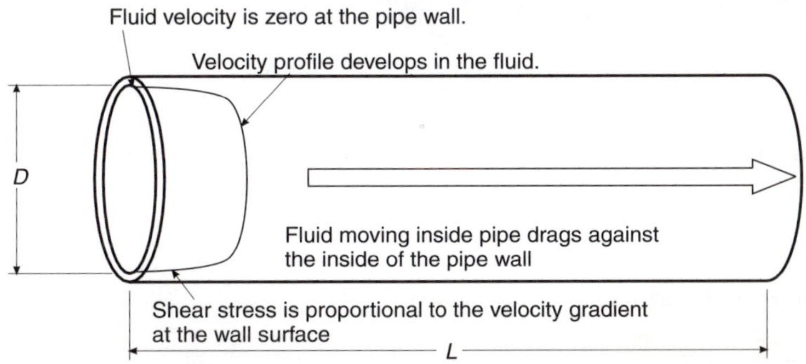

Figure 2.1 Origin of the force that is exerted by the flowing fluid on the pipe

fluids that flow in a slow orderly manner that is barely turbulent inside a smooth pipe to values that are only about one tenth of that value when the fluid moves very fast inside the pipe and is intensely turbulent.

The average velocity of the fluid is related to the total volumetric flowrate Q by

$$\bar{V} = \frac{Q}{\frac{\pi}{4}D^2} \text{ m/s} \tag{2.3}$$

Substitution of equation 2.2 into equation 2.1 gives

$$\text{Force} = \pi DL\left(\frac{1}{2}\rho_f \bar{V}^2\right)f \text{ N} \tag{2.4}$$

The force is generated by the pressure gradient along the pipe. When the fluid is flowing steadily equation 2.4 can be converted into a form that gives the pressure gradient due to friction (*PGDTF*) as the fluid flows through the pipe under steady conditions.

$$\begin{aligned} PGDTF &= -\frac{\Delta P_f}{L} = \frac{\text{Force}}{\frac{\pi}{4}D^2L} = \frac{2}{D}\rho_f \bar{V}^2 f \\ &= \frac{4\tau_w}{D} \text{ N/m}^2 \end{aligned} \tag{2.5}$$

The symbol ΔP_f in equation 2.5 represents the pressure drop that the flowing fluid experiences due only to the fractional drag on the pipe wall. The pressure decreases in the direction of flow so that ΔP_f has a negative numerical value. This makes *PGDTF* a positive quantity. Equation 2.5 provides a method for the experimental determination of the friction factor, f, because *PGDTF* can be measured in the laboratory.

The energy dissipated by the frictional drag can be calculated from the force exerted by the fluid on the pipe wall. The energy dissipation is calculated as energy used per unit mass of fluid.

$$F = \frac{\text{Force} \times L}{\frac{\pi}{4}D^2 L \rho_f} \text{ J/kg} \tag{2.6}$$

Using equations 2.4 and 2.5 this becomes

$$F = 2f\frac{\bar{V}^2 L}{D} = \frac{-\Delta P_f}{\rho_f} \text{ J/kg} \tag{2.7}$$

It is common practice to express pressure in terms of the equivalent height of a column of fluid in the gravitational field of the earth.

$$-\Delta P_f = g\rho_f h_f \tag{2.8}$$

h_f is called the head loss due to friction

$$h_f = 2\frac{f\bar{V}^2}{g}\frac{L}{D} \tag{2.9}$$

The head loss due to friction is sometimes also expressed in terms of the number of velocity heads, N_{vh}, that are lost. A velocity head is defined to be the quantity $\bar{V}^2/2g$ so that

$$h_f = N_{vh}\frac{\bar{V}^2}{2g} \tag{2.10}$$

Comparing this with equation 2.9

$$N_{vh} = 4f\frac{L}{D} \tag{2.11}$$

Whenever the pipe has a cross-section that is not circular but the pipe still runs full, the diameter D in all of the above formulas should be replaced by the hydraulic mean diameter which is defined by

$$D_H = \frac{4 \times \text{flow cross-sectional area}}{\text{wetted perimeter}} \tag{2.12}$$

2.2 The friction factor

Experiments have shown that the friction factor can be correlated uniquely with the Reynolds number calculated for the fluid as it flows inside the channel.

$$Re = \frac{D\bar{V}\rho_f}{\mu_f} \tag{2.13}$$

A large amount of data obtained experimentally using many different fluids in pipes having diameters differing by orders of magnitude have been assembled into the so-called friction-factor chart. This chart is shown in Figure 2.2 and it is probably the most widely used chart by engineers who deal with fluid flow problems. It is important to understand that the friction factor plot represents experimentally determined data and does not have *a priori* theoretical foundation. It appears in virtually every text book that covers hydrodynamics and fluid flow. It has been published in several sizes and prior to the personal computer era it was common to read values directly from the graph with the attendant lack of precision. Now computer versions of the friction factor chart are readily available and the chart is included in the FLUIDS toolbox that is included on the CD-ROM that accompanies this book.

Empirical expressions have been established that summarize these graphs and the most widely used equation is

$$\frac{1}{\sqrt{f}} = 1.74\ln(Re\sqrt{f}) - 0.40 \tag{2.14}$$

Equation 2.14 applies only when the inside of the pipe wall is smooth and the fluid is turbulent in the pipe which occurs when $Re > 2000$.

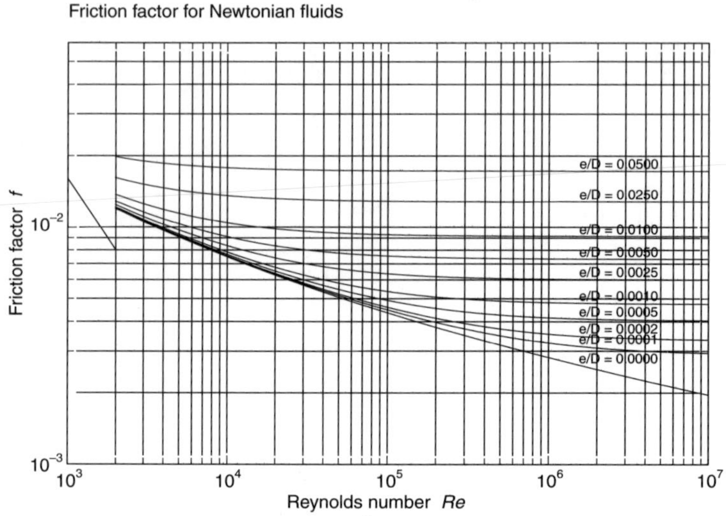

Figure 2.2 Friction factor plotted against the pipe Reynolds number. These graphs were generated using equation 2.15

When the inside of the pipe wall is rough the friction factor increases and equation 2.14 is modified to

$$\frac{1}{\sqrt{f}} = -1.74\ln\left(0.338\frac{e}{D} + \frac{1}{Re\sqrt{f}}\right) - 0.4$$
$$\frac{1}{\sqrt{f}} = -1.74\ln\left(0.27\frac{e}{D} + \frac{1.25}{Re\sqrt{f}}\right) \tag{2.15}$$

e/D is a measure of the surface roughness relative to the pipe diameter. *e* is the average height of any rough features on the inner surface. Typical values of the surface roughness for some common materials are given in Table 2.1. Equation 2.15 is commonly referred to as the Colebrook equation.

Equations 2.14 and 2.15 are not particularly convenient to use because neither gives the value of *f* as an explicit function of the Reynolds number, *Re*.

An approximate formula for the friction factor over a restricted range of Reynolds number that is often used because it does give *f* as an explicit function of *Re*, is the Blasius equation

$$f = 0.079\,Re^{-0.25} \tag{2.16}$$

This equation is a reasonably good representation of the data for smooth pipes over the range $2000 < Re < 100\,000$.

When $Re < 2000$ the flow is laminar and the friction factor is given by

$$f = \frac{16}{Re} \tag{2.17}$$

This relationship is derived in Section 5.2.1

Table 2.1 Effective roughness of various surfaces (Source: Darby, 1996)

Material	*Condition*	*Roughness range* (mm)	*Recommended value* (mm)
Drawn copper, brass or stainless steel	New	0.0015–0.01	0.002
Commercial steel	New	0.02–0.1	0.045
	Light rust	0.15–1.0	0.3
	General rust	1–3	2.0
Iron	Wrought, new	0.045	0.045
	Cast, new	0.25–1	0.3
	Galvanized	0.025–0.15	0.15
	Asphalt-coated	0.1–1.0	0.15
Sheet metal	Ducts, smooth joints	0.02–0.1	0.03
Concrete	Very smooth	0.025–0.18	0.04
	Wood floated, brushed	0.2–0.8	0.3
	Rough, visible form marks	0.8–2.5	2.0
Wood	Stave	0.25–1.0	0.5
Glass and plastic	Drawn tubing	0.0015–0.01	0.002
Rubber	Smooth tubing	0.006–0.07	0.01
	Wire-reinforced	0.3–4.0	1.0

2.3 Calculation of pressure gradient and flowrate

It should be clear that the correlations for the friction factor f given in the friction-factor chart make it possible to calculate the pressure gradient in a pipe whenever the flowrate and the pipe diameter are known. It is simply a matter of calculating the Reynolds number Re and reading the corresponding friction factor on the chart. The corresponding pressure gradient due to friction can be calculated using Equation 2.5. Reading values from the friction-factor chart is not particularly accurate but accurate values can be easily and conveniently obtained from the FLUIDS software package on the CD-ROM that accompanies this book.

Illustrative example 2.1

Calculate the pressure gradient due to friction when water flows through a smooth 10 cm diameter pipe at 1.5 m/s. Assume this data for the water: $\rho_w = 1000\,\text{kg/m}^3$ and $\mu_f = 0.001\,\text{kg/ms}$.

$$\begin{aligned} Re &= \frac{D\bar{V}\rho_f}{\mu_f} \\ &= \frac{0.10 \times 1.5 \times 1000}{0.001} \\ &= 1.5 \times 10^5 \end{aligned}$$

The friction factor plot can be read from the friction-factor chart or preferably obtained from the FLUIDS toolbox using the single-phase fluid friction

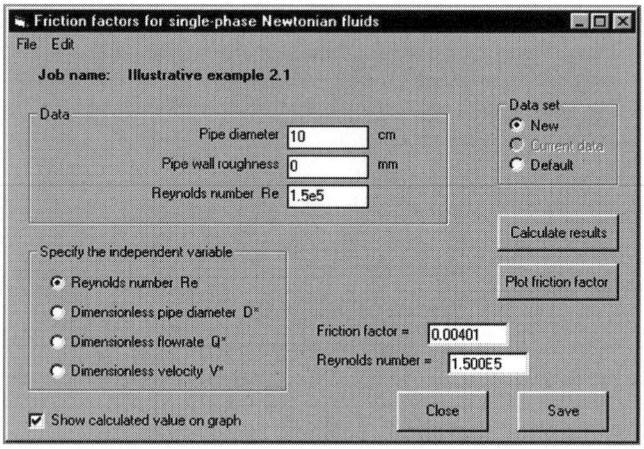

Figure 2.3 Data input screen to calculate friction factor using the FLUIDS toolbox

factor screen as shown in Figure 2.3. The value for f at the specified value of Re is 0.0040.

The pressure gradient due to friction is obtained from equation 2.5

$$PGDTF = \frac{-\Delta P_f}{L} = \frac{2\rho_f \bar{V}^2 f}{D}$$

$$= \frac{2 \times 1000 \times 1.5^2 \times 0.00401}{0.1}$$

$$= 180.4\,\text{Pa/m}$$

Illustrative example 2.2

Calculate the increase in the pressure gradient due to friction if the inside of the pipe wall has roughness = 0.1 mm.

This can be done conveniently by changing the 'Pipe wall roughness' entry on the data form as shown in Figure 2.4 and calculating the new value of the friction factor which is $f = 0.00518$ as shown.

The new value of *PGDTF* is

$$PGDTF = \frac{2 \times 1000 \times 1.5^2 \times 0.00518}{0.1} = 233.1\,\text{Pa/m}$$

Note the significant increase in pressure gradient due to this comparatively small increment in roughness of the wall surface.

The conventional friction-factor chart is not at all convenient for the calculation of the flowrate when the available pressure gradient and the pipe diameter are known and an alternative method is developed here that facilitates calculations of this type.

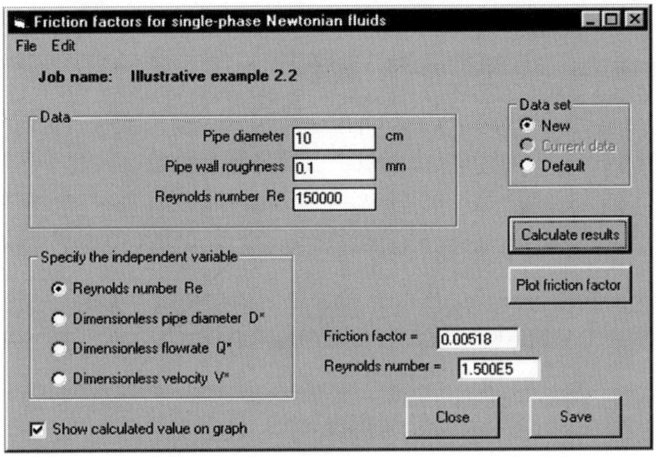

Figure 2.4 Specification of parameters for illustrative example 2.2

The relationship between f and Re shown in Figure 2.2 is plotted as a series of graphs of f against $(Re^2 f)^{1/3}$ as shown in Figure 2.5. An explicit method for the calculation of the flowrate through a pipe when the diameter and available pressure gradient are known can be readily developed using this plot. When the fluid is flowing steadily the friction factor is related to the pressure gradient by equation 2.5 which gives

$$\bar{V}^2 f = \frac{PGDTF \times D}{2\rho_f} \tag{2.18}$$

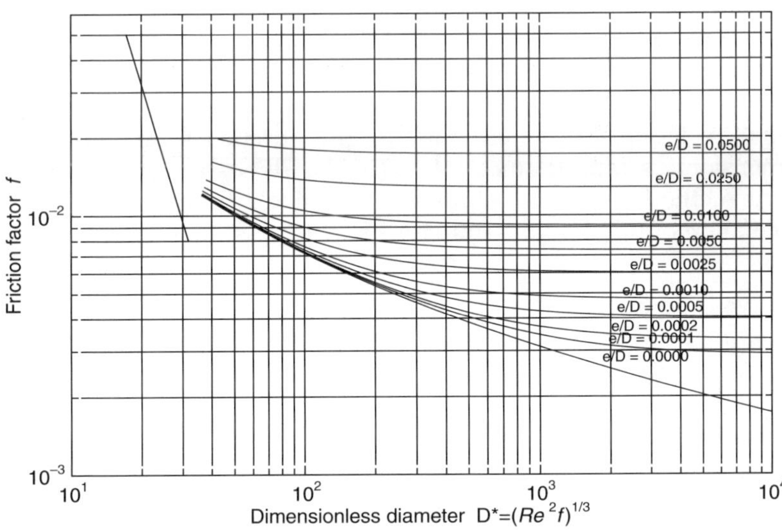

Figure 2.5 Friction factor plotted against the dimensionless pipe diameter. Use this chart if the pipe diameter and the *PGDTF* are known

Thus

$$Re^2 f = \frac{D^2 \bar{V}^2 \rho_f^2 f}{\mu_f^2}$$
$$= \frac{D^3 \rho_f \times PGDTF}{2\mu_f^2} = D^{*3} \tag{2.19}$$

D^* is called the dimensionless pipe diameter and it can be calculated independently of the fluid velocity. Once D^* is known the friction factor can be calculated explicitly from equation 2.15 or read from Figure 2.5

The Colebrook equation 2.15 can be written in terms of the dimensionless pipe diameter D^*

$$\frac{1}{\sqrt{f}} = -1.768 \ln\left(0.27\frac{e}{D} + \frac{1.25}{D^{*3/2}}\right) \tag{2.20}$$

This gives an explicit equation for the friction factor f. Once f and D^* are known the Reynolds number can be calculated from

$$Re = D^{*3/2} \times \frac{1}{\sqrt{f}} \tag{2.21}$$

This solution can be accepted only if $Re > 2000$ because equation 2.15 applies only to the turbulent flow region.

When $Re < 2000$ the flow is laminar and the relationship between f and D^* can be developed by combining equations 2.17 and 2.19 to give

$$f = \frac{16^2}{D^{*3}} \tag{2.22}$$

This plots as a straight line having slope = −3 in the logarithmic coordinate system.

This method can be used conveniently by reading the friction factor directly from Figure 2.5 which is the original friction factor plotted against the dimensionless pipe diameter D^*. The FLUIDS software toolbox implements the method directly and provides a particularly convenient method for the calculation of the flowrate under a known pressure gradient.

Illustrative example 2.3

Calculate the flowrate that will be achieved when water is forced through a smooth 10-cm pipe under a pressure gradient of 180 Pa/m.

$$D^{*3} = \frac{D^3 \rho_f PGDTF}{2\mu_f^2}$$
$$= \frac{0.1^3 \times 1000 \times 180}{2 \times 0.001^2} = 9.0 \times 10^7$$
$$D^* = 448.1$$

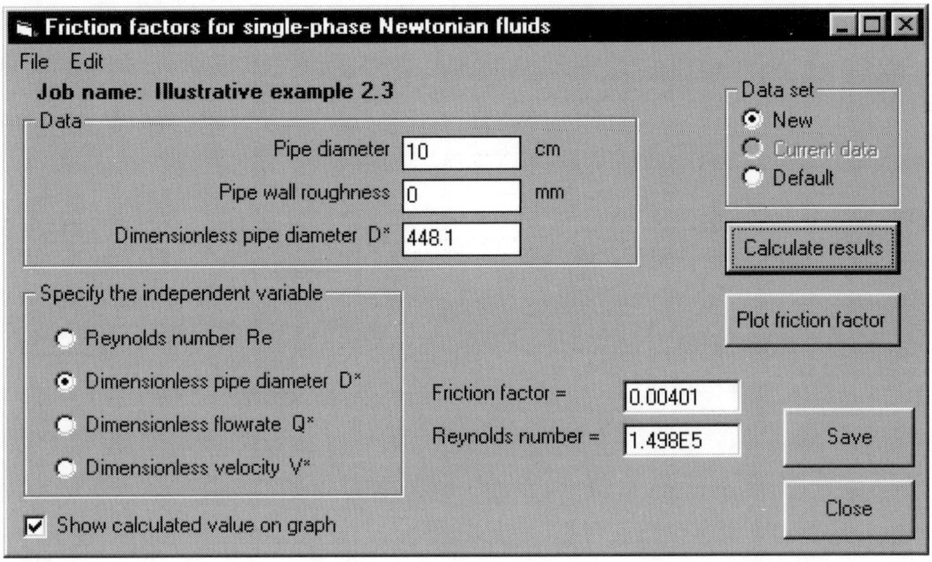

Figure 2.6 Specification of toolbox parameters for illustrative example 2.3

Use Figure 2.5 or the FLUIDS toolbox (see Figure 2.6) or use equation 2.20 directly:

$$\begin{aligned}
\frac{1}{\sqrt{f}} &= -1.768 \ln\left(0 + \frac{1.252}{D^{*3/2}}\right) \\
&= 15.79 \\
f &= 0.00401 \\
Re &= D^{*3/2} \times \frac{1}{\sqrt{f}} = 9.487 \times 10^3 \times 15.79 \\
&= 1.5 \times 10^5 \\
\bar{V} &= \frac{Re\mu_f}{D\rho_f} = \frac{1.5 \times 10^5 \times 0.001}{0.1 \times 1000} = 1.5\,\text{m/s} \\
Q &= \frac{\pi}{4} D^2 \bar{V} = \frac{\pi\, 0.1^2 1.5}{4} = 1.178 \times 10^{-2}\,\text{m}^3/\text{s}
\end{aligned}$$

Other dimensionless groups can be used to develop similar direct calculation methods for different problem situations. When the diameter of the pipe must be selected to transport a fluid at a specified flow rate $Q\,\text{m}^3/\text{s}$ under a given frictional pressure loss, the dimensionless group $Re^5 f$ is useful. This is related to the dimensionless flowrate which is denoted by Q^*

$$Q^{*3} = Re^5 f \tag{2.23}$$

The friction factor is plotted against Q^* in Figure 2.7.

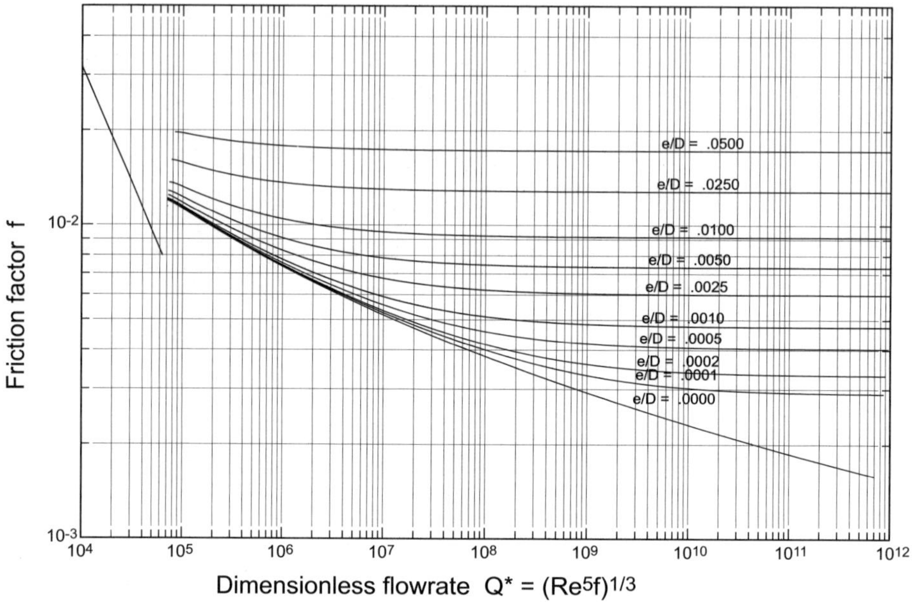

Figure 2.7 Friction factor plotted against the dimensionless volumetric flowrate. Use this chart if the volumetric flowrate and the *PGDTF* are known

When the fluid is flowing steadily in the pipe

$$\begin{aligned} Re^5 f &= Re^3\ Re^2 f \\ &= Re^3\ D^{*3} \\ &= \frac{D^3 \bar{V}^3 \rho_f^3}{\mu_f^3} \times \frac{D^3 \rho_f PGDTF}{2\mu_f^2} \end{aligned} \tag{2.24}$$

where equation 2.19 has been used.

The velocity, $\bar{V}$, is related to the flow rate and the pipe diameter by

$$\bar{V} = \frac{Q}{\frac{\pi}{4} D^2} \tag{2.25}$$

and substituting this in equation 2.24

$$Q^{*3} = Re^5 f = \frac{\rho_f^4 D^6 \bar{V}^3 PGDTF}{2\mu_f^5} = \frac{32 \rho_f^4 PGDTF}{\pi^3 \mu_f^5} Q^3 \tag{2.26}$$

Q^* can be calculated without requiring the pipe diameter or the average velocity to be known. The use of Q^* for practical problem solving is described in the following illustrative example.

Illustrative example 2.4

Calculate the diameter of a smooth pipe that would transport $0.01178\,\text{m}^3/\text{s}$ of water under a pressure gradient of 180.0 Pa/m (see Figure 2.8).

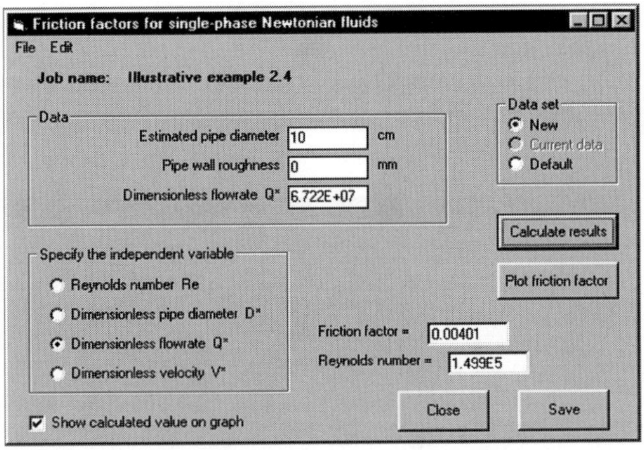

Figure 2.8 Data input screen to calculate friction factor for illustrative example 2.4

$$
\begin{aligned}
Q^{*3} &= \frac{32\rho_f^4 PGDTF Q^3}{\pi^3 \mu_f^5} \\
&= \frac{32 \times 1000^4 \times 180.0 \times 0.01178^3}{\pi^3 \times 0.001^5} \\
&= 3.037 \times 10^{23} \\
Q^* &- 6.722 \times 10^7
\end{aligned}
$$

From Figure 2.7, $f = 0.004$

$$
\begin{aligned}
Re^5 &= \frac{Q^{*3}}{f} = \frac{3.037 \times 10^{23}}{0.004} = 7.593 \times 10^{25} \\
Re &= (7.593 \times 10^{25})^{1/5} = 1.5 \times 10^5 \\
Re &= \frac{D\bar{V}\rho_f}{\mu_f} = \frac{4Q\rho_f}{\pi D \mu_f} \\
D &= \frac{4Q\rho_f}{\pi Re \mu_f} = \frac{4 \times 0.01178 \times 1000}{\pi \times 1.5 \times 10^5 \times 0.001} \\
&= 0.1\,\text{m}
\end{aligned}
$$

When the pipe diameter must be selected to transport a fluid at specified velocity with a given frictional pressure loss, the following dimensionless group is useful

$$
V^{*3} = \frac{Re}{f} = \frac{2\rho_f^2}{\mu_f} \frac{1}{PGDTF} \bar{V}^3 \tag{2.27}
$$

where f has been substituted using equation 2.5. V^* is called the dimensionless velocity and it can be evaluated without requiring the pipe diameter or the total flowrate to be known. The friction factor is plotted against V^* in Figure 2.10. The use of V^* is demonstrated in the following illustrative example.

Illustrative example 2.5

Calculate the diameter of a smooth pipe that would transport water at 1.5 m/s under a pressure gradient of 180 Pa/m. What is the volumetric flowrate? (see Figure 2.9.)

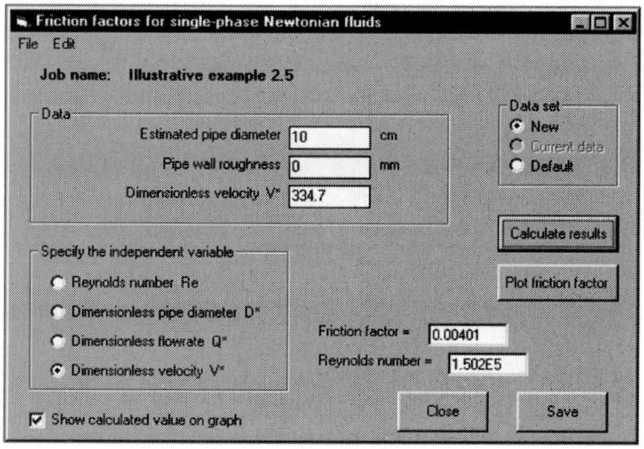

Figure 2.9 Data input screen to calculate friction factor for illustrative example 2.5

$$
\begin{aligned}
V^{*3} &= \frac{2\rho_f^2 \bar{V}^3}{\mu_f PGDTF} \\
&= \frac{2 \times 1000^2 \times 1.5^3}{0.001 \times 180} \\
&= 3.750 \times 10^7 \\
V^* &= 3.347 \times 10^2
\end{aligned}
$$

From Figure 2.10, $f = 0.004$.

$$
\begin{aligned}
fRe &= V^{*3} f \\
&= 3.750 \times 10^7 \times 0.004 \\
&= 1.5 \times 10^5 \\
D &= \frac{Re\ \mu_f}{\bar{V}\rho_f} \\
&= \frac{1.5 \times 10^5}{1.5 \times 1000} = 0.1\,\text{m} \\
Q &= \frac{\pi}{4} D^2 \bar{V} = \frac{\pi \times 0.1^2 \times 1.5}{4} = 0.01178\,\text{m}^2/\text{s}
\end{aligned}
$$

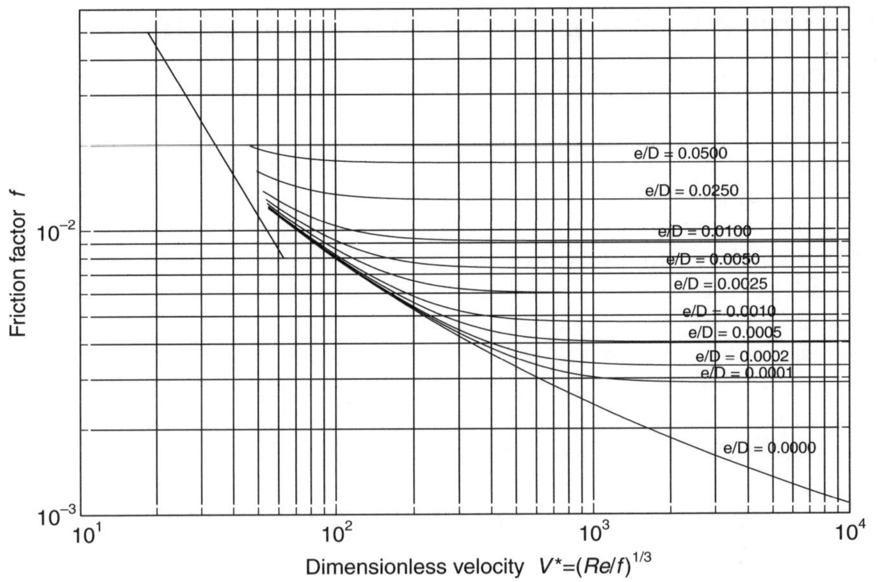

Figure 2.10 Friction factor plotted against the dimensionless fluid velocity. Use this plot if the required velocity and *PGDTF* are known

When plotting each of the four different versions of the friction factor plot, each of the groups D^*, V^* and Q^* is evaluated at any value of *Re* using the data from the original friction factor plot (Figure 2.2) or from equation 2.15. Each of these charts is nothing more than a recalculation of the information contained in Figure 2.2. They are easy to generate using any standard spreadsheet program by starting with the two columns containing *Re* and *f* pairs calculated from the Colebrook equation. Columns can easily be generated in the spreadsheet for the quantities $(Re^2 f)^{1/3}$, $(Re^5 f)^{1/3}$, and $(Re/f)^{1/3}$ using data from the original columns and plotted as needed. The choice of which form of the chart to use depends entirely on the data that is available for the problem on hand – if the frictional pressure loss must be calculated then the original friction factor chart (Figure 2.2) is appropriate. If the pressure gradient due to friction is known or specified then Figures 2.5, 2.7, or 2.10 is used depending on which of the variables D, $\bar{V}$, or Q is known. The use of these charts obviates the need to use iterative or trial-and-error calculations to solve practical problems. The only exception is when the pipe diameter must be calculated for rough materials. Then the relative roughness must be estimated before the friction factor can be read from Figure 2.7 or 2.10 and the calculation must be refined after the first estimate of the relative roughness has been obtained. The four versions of the friction factor plot can be conveniently generated using the friction factor button on the main menu of the FLUIDS software toolbox.

2.4 The energy balance for piping systems

When a fluid moves through any constrained space, it will transfer some of its momentum to the solid boundaries and in so doing will dissipate energy. The transfer of momentum will manifest itself as a force on any solid surface that is in contact with the moving fluid. The energy dissipation usually manifests itself by noting that an energy source is required to keep the fluid moving. The energy source can be a pump, the potential energy in the gravitational field that causes rivers to run downhill, or the thermal radiant energy that is captured by the surface of warm oceans and drives tropical atmospheric storms and hurricanes. In situations where the fluid must be transported, it is necessary to provide sufficient energy to the fluid to make good the dissipated energy as well as to provide any potential or kinetic energy that is required. A procedure for setting up the overall energy balance for a flowing fluid is established in this section.

2.4.1 Internal energy and energy dissipation in a flowing fluid

The energy that is contained within the fluid mass by virtue of thermal and other motions of its molecules is the internal energy. Any change in the physical condition of the fluid (an increase in temperature for example) will give rise to a change in internal energy. Although changes in internal energy are quite small (especially for incompressible fluids) they must be accounted for in the overall energy balance.

The energy that is dissipated by frictional losses against solid bounding surfaces can be quantified using the concept of entropy. Frictional dissipation is essentially an irreversible process so that the entropy change associated with unit mass of flowing fluid will reflect the energy that is degraded into heat. If the flow process were perfectly reversible and isothermal and no energy degraded into heat, $T\Delta s$ would be exactly equal to q which is the energy that is transferred across the bounding wall by the thermal transfer processes. However, degradation of energy by friction effectively adds more heat to the system by an amount of F.

$$T\Delta s = q + F \tag{2.28}$$

F can be evaluated using the simple ideas that were developed in Section 2.1 using a straight section of cylindrical pipe as the physical model. The energy dissipated by frictional drag on the inside surface of the pipe wall is evaluated as the force exerted on the fluid multiplied by the distance over which the fluid slides.

$$\begin{aligned} \text{Force} &= \pi D \tau_{\text{w}} L \\ \text{Distance} &= L \\ \text{Mass of fluid} &= \frac{\pi}{4} D^2 L \rho_{\text{f}} \end{aligned} \tag{2.29}$$

Energy dissipated per unit mass of fluid, F is given by

$$\begin{aligned} F &= \frac{\pi D \tau_w L^2}{\frac{\pi}{4} D^2 L \rho_f} = \frac{4\tau_w L}{\rho_f D} \\ &= 2f\bar{V}^2 \frac{L}{D} \end{aligned} \tag{2.30}$$

2.4.2 The flow energy

A separate quantity of energy must be accounted for just to keep the fluid moving. This concept is illustrated in Figure 2.11 by means of a conceptual and general system into which fluid flows at pressure P_1 and from which fluid leaves at pressure P_2. The system is considered to be at steady state.

Work done on the system to get 1 kg of fluid in at P_1

$$= P_1 \times \frac{\pi}{4} D_1^2 \times \Delta x_1 = P_1 v_1 \tag{2.31}$$

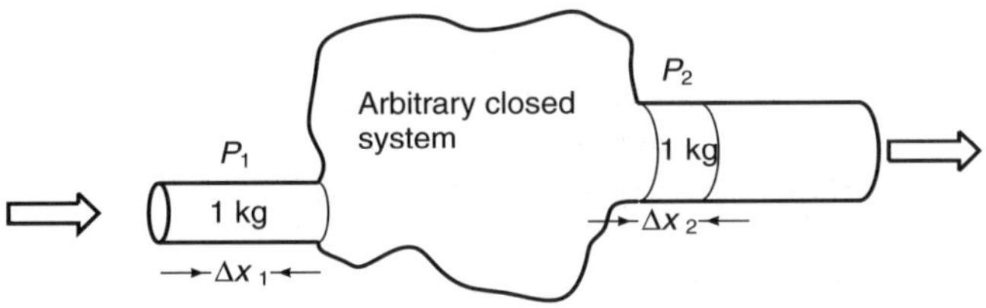

Figure 2.11 Schematic representation of the flow energy

v_1 is the specific volume of the fluid at the inlet.

Work done by the system in forcing 1 kg of fluid out at P_2

$$= P_2 \times \frac{\pi}{4} D_2^2 \times \Delta x_2 = P_2 v_2 \tag{2.32}$$

The net work done by the system just to get the fluid in and out is

$$P_2 v_2 - P_1 v_1 = \Delta P v \tag{2.33}$$

2.4.3 Potential and kinetic energy

The potential energy of 1 m^3 of fluid that is situated at an elevation z above an appropriate datum is given by

$$\text{PE} = g\rho_f z \ \text{J/m}^3 \tag{2.34}$$

The potential energy per unit mass is

$$\text{PE} = gz \ \text{J/kg} \tag{2.35}$$

The kinetic energy of 1 kg of fluid that is moving at $\bar{V}$ m/s is

$$\text{KE} = \frac{1}{2}\bar{V}^2 \ \text{J/kg} \tag{2.36}$$

2.4.4 The overall energy balance

When a fluid is moved from one location to another, such as when it is pumped through a piping system, there is usually a redistribution of energy. For example when a fluid flows downward in the gravitational field of the earth, potential energy can be converted to kinetic energy and the fluid can gain energy from a pump. Electrical energy consumed by a pump motor can be converted to kinetic or potential energy of the fluid.

As the fluid moves the changes in each type of energy must be balanced against the net loss or gain of energy by the fluid. An overall energy balance that accounts for all types of energy that influences the behavior of fluids that are of interest in typical engineering situations is given by

$$\Delta u + \Delta Pv + g\Delta z + \frac{1}{2}\Delta \bar{V}^2 = q - W \tag{2.37}$$

In this equation the symbol Δ indicates the change in the specified energy component as the fluid changes its location. All energies are expressed per unit mass of fluid.

The internal energy is determined largely by the temperature of the fluid and this is not of great consequence in situations that are of interest here. The term q which represents energy exchange with the environment by thermal transfer mechanisms is also not required specifically in situations where slurries and pure fluids are moved in piping systems. As a result a slightly different formulation of the overall energy balance is developed by making use of two thermodynamic relationships.

$$\Delta u = T\Delta s - P\Delta v \tag{2.38}$$

and

$$T\Delta s = q + F \tag{2.39}$$

substitution using $\Delta Pv = v\Delta P + P\Delta v$ gives

$$W + F + v\Delta P + g\Delta z + \frac{1}{2}\Delta \bar{V}^2 = 0 \tag{2.40}$$

Because the terms that reflect thermal energy have been eliminated in equation 2.40, this equation is often referred to as the mechanical energy balance for flowing fluids.

2.5 The effect of fittings in a pipeline

Valves and fittings add friction to the line because of increased turbulence that is induced in the fluid as it passes through the fitting. Several different methods are available to estimate the additional energy dissipation. Some common fittings are shown in Figure 2.12.

The simplest method is based on the assumption that a particular fitting will contribute to the energy dissipation an amount equivalent to an

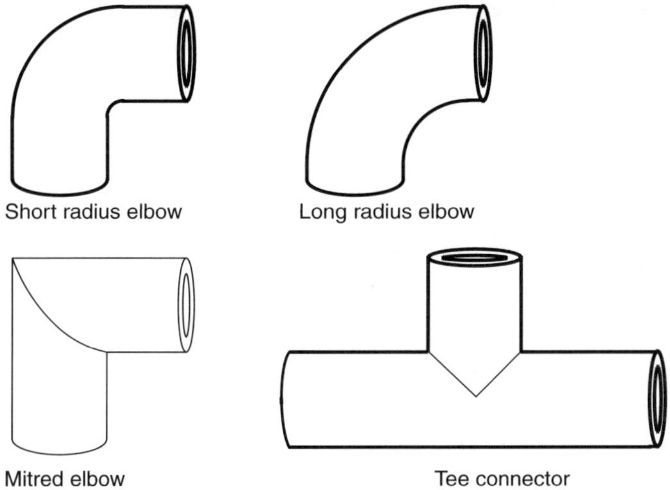

Figure 2.12 Common pipe fittings that contribute additional energy dissipation in a pipeline

additional length of pipe. The additional length is calculated as a multiple of the pipe diameter. The number of pipe diameters contributed by each type of fitting is given in Table 2.2. The energy dissipation is given by equation 2.13

$$F = 2f\bar{V}^2 \frac{L}{D} \tag{2.41}$$

where *L/D* represents the number of pipe diameters contributed by the fitting.

An alternative method is based on the idea that kinetic energy is dissipated as the fluid flows through the fitting and the loss is calculated in terms of the number of velocity heads that are lost.

Table 2.2 Friction losses in pipe fittings

	Number of pipe diameters	*Number of velocity heads $\bar{V}^2/2g$*
45° elbows	15	0.3
90° elbows (standard radius)	35	0.7
90° square elbows	60	1.2
Exit from leg of T-piece	60	1.2
Entry into leg of T-piece	90	1.8
Unions and couplings	Small	Small
Globe valves fully open	60–300	1.2–6.0
Gate valves: fully open	7	0.15
$^3/_4$ open	40	1
$^1/_2$ open	200	4
$^1/_4$ open	800	16

The frictional head loss in the fitting is given in terms of velocity heads by equation 2.10

$$h_f = N_{vh}\frac{\bar{V}^2}{2g}$$
$$F = gh_f = N_{vh}\frac{\bar{V}^2}{2} \tag{2.42}$$

If there is a change in velocity through the fitting, it is the upstream velocity that is used to calculate the value of the velocity head. The frictional dissipation for a variety of common fittings is given in terms of the equivalent additional length of pipe and in terms of the number of lost velocity heads in Table 2.2.

A more accurate method is based on experimental data that was obtained in a variety of fittings and is correlated in terms of a loss coefficient K_f.

$$F = \frac{K_f\bar{V}^2}{2} \tag{2.43}$$

The loss coefficient is related to the upstream Reynolds number of the flow through the fitting

$$K_f = \frac{K_1}{Re} + K_\infty\left(1 + \frac{0.0254}{D}\right) \tag{2.44}$$

The values of K_1 and K_∞ are given in Table 2.3 for a variety of fittings, in Table 2.4 for pipe expansions and contractions, and in Table 2.5 for entrances and

Table 2.3 *K* Constants for Loss Coefficients for Valves and Fittings. To be used with equation 2.44

		Fitting		K_1	K_∞
Elbows	90°	Standard (($r/D = 1$)	Threaded	800	0.40
		Standard ($r/D = 1$)	Flanged/Welded	800	0.25
		Long Radius ($r/D = 1.5$)	All types	800	0.20
		Mitered	One weld(90°)	1000	1.15
			Two welds(45°)	800	0.35
			Three welds(30°)	800	0.30
			Four welds(22.5°)	800	0.27
			Five welds(18°)	800	0.25
	45°	Standard ($r/D = 1$)	All types	500	0.20
		Long Radius ($r/D = 1.5$)		500	0.15
		Mitered	One weld (45°)	500	0.25
		Mitered	Two welds (22.5°)	500	0.15
	180°	Standard ($r/D = 1$)	Threaded	1000	0.70
		Standard ($r/D = 1$)	Flanged/Welded	1000	0.35
		Long radius ($r/D = 1.5$)	All types	1000	0.30

Table 2.3 *Continued.*

		Fitting		K_1	K_∞
Tees	Used as elbow	Standard	Threaded	500	0.70
		Long Radius	Threaded	800	0.40
		Standard	Flanged/Welded	800	0.80
		Stub-in Branch		1000	1.00
	Run-through	Threaded		200	1.10
		Flanged/Welded		150	0.05
		Stub-in Branch		100	0.00
Valves (Fully open)	Gate, ball or plug	Full Line Size, $\beta = 1.0$		300	0.10
		Reduced Trim, $\beta = 0.9$		500	0.15
		Reduced Trim, $\beta = 0.8$		1000	0.25
		Standard		1500	4.00
	Globe	Angle or Y-type		1000	2.00
	Diaphragm	Dam type		1000	2.00
	Butterfly			800	0.25
	Check	Lift		2000	10.00
		Swing		1500	1.50
		Tilting disk		1000	0.50

Source: Hooper (1981).

Table 2.4 Loss Coefficients for Pipe Expansions and Contractions

	K_f to be used with upstream velocity head, $V_1^2/2$ $\beta = D_1/D_2$	
Contraction		
$\theta < 45°$	$Re_1 < 2500$	$K_f = 1.6(1.2 + \frac{160}{Re_1})(\beta^4 - 1)\sin(\theta/2)$
	$Re_1 > 2500$	$K_f = 1.6(0.6 + 1.92f_1)\beta^2(\beta^2 - 1)\sin(\theta/2)$
$45° \le \theta \le 180°$	$Re_1 < 2500$	$K_f = (1.2 + \frac{160}{Re_1})(\beta^4 - 1)(\sin(\theta/2))^{0.5}$
	$Re_1 > 2500$	$K_f = (0.6 + 1.92f_1)\beta^2(\beta^2 - 1)(\sin(\theta/2))^{0.5}$
Expansion		
$\theta < 45°$	$Re_1 < 4000$	$K_f = 5.2(1 - \beta^4)\sin(\theta/2)$
	$Re_1 > 4000$	$K_f = 2.6(1 + 3.2f_1)(1 - \beta^2)^2\sin(\theta/2)$
$45° \le \theta \le 180°$	$Re_1 < 4000$	$K_f = 2(1 - \beta^4)$
	$Re_1 > 4000$	$K_f = (1 + 3.2f_1)(1 - \beta^2)^2$

Re_1 is the upstream Reynolds number, and f_1 is the pipe friction factor at this Reynolds number.
Source: Hooper (1988).

Table 2.5 Loss Coefficients for Pipe Entrance and Exit

$K_f = K_1/N_{Re} + K_\infty$		
Entrance to pipe		
Inward projecting (Borda)	$K_1 = 160$	$K_\infty = 1.0$
Flush	$K_1 = 160$	$K_\infty = 0.5$
Entrance to tank	$K_1 = 0.0$	$K_\infty = 1.0$ for all geometries

Source: Hooper (1981) Published by permission.

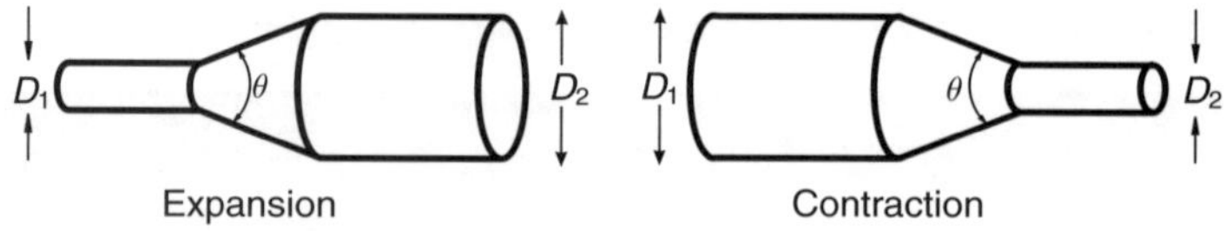

Figure 2.13 Pipe expansion and contraction

exits from tanks. Comparison of equation 2.43 with equation 2.42 shows that K_f is equivalent to the number of velocity heads that are lost as the fluid passes through the fitting.

The geometry of a pipe expansion and contraction is shown in Figure 2.13.

Illustrative example 2.6

Water is drained from the upper open reservoir as shown in Figure 2.14 through a 5 cm steel pipe having a wall roughness of 0.05 mm. Calculate the flowrate that will be achieved in the pipeline.

This simple problem requires the velocity of the water in the pipeline to be determined. A convenient trial and error method can be developed easily using a spreadsheet.

The frictional dissipation of energy in the fittings can be calculated using equation 2.43

$$F_{\text{fitting}} = \frac{K_f \bar{V}^2}{2}$$

For the pipe entrance $\dfrac{160}{Re} + 0.5$

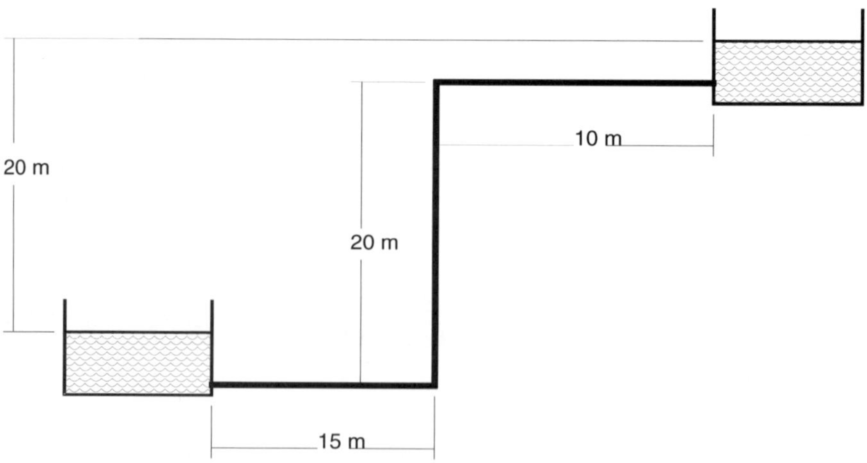

Figure 2.14 Simple transfer pipeline

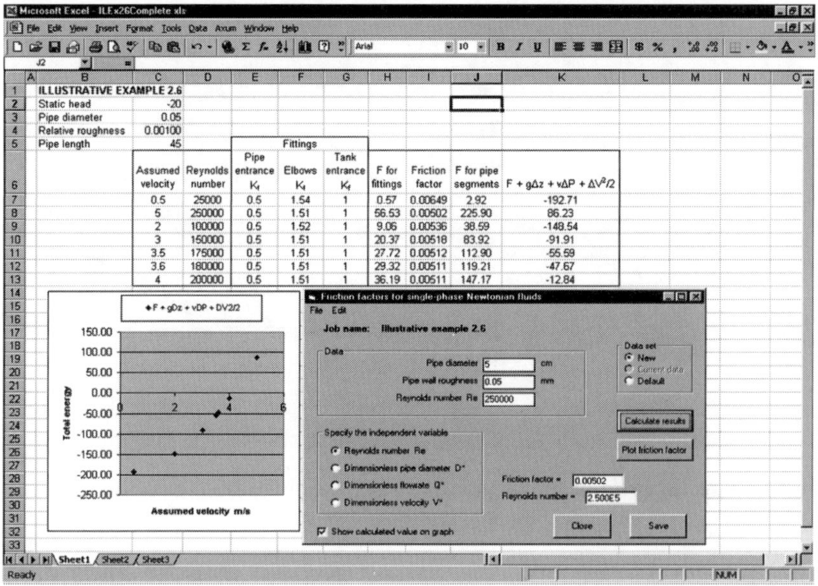

Figure 2.15 The solution is facilitated by using a spreadsheet together with the friction factor module in the FLUIDS toolbox

For the elbows:

$$K_f = \frac{800}{Re} + 0.40\left(1 + \frac{0.0254}{0.05}\right) = \frac{800}{Re} + 1.508$$

For the tank entrance: $K_f = 1.0$
The frictional dissipation of energy in the pipe is given by

$$F = 2f\bar{V}^2\frac{L}{D}$$

A spreadsheet can be set up to calculate these quantities from an assumed velocity $\bar{V}$ and therefore the Reynolds number. A typical spreadsheet format is shown in Figure 2.15. In order to calculate F, the friction factor must be determined. This can be done most conveniently by calling up the friction factor module in the FLUIDS toolbox. This will run in a separate window as shown in Figure 2.15, and it is a simple matter to transfer the Reynolds number from any row in the spreadsheet to the FLUIDS window and return the calculated friction factor to the spreadsheet. In this way, various values of $\bar{V}$ can be tried until the energy balance is satisfied.

$$F + W + g\Delta z + v\Delta P + \frac{\Delta\bar{V}^2}{2} = 0$$

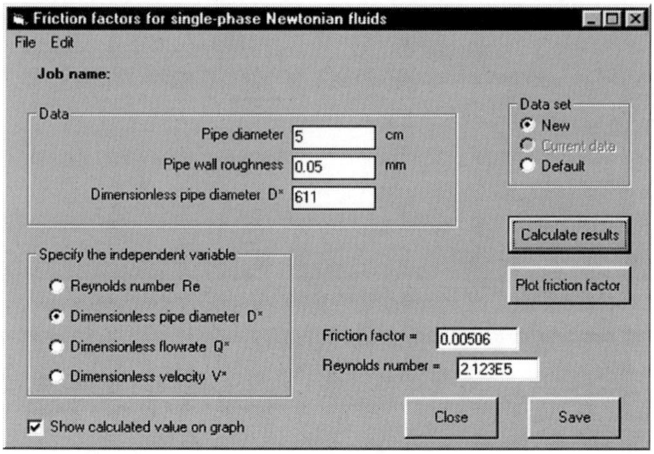

Figure 2.16 Specification of toolbox parameters for illustrative example 2.6

In this problem W, ΔP and $\Delta\bar{V}$ are all zero so that the energy balance becomes

$$F + g\Delta z = 0$$

A simple graph allows the water velocity to be established easily as

$$\bar{V} = 4.1\,\mathrm{m/s}$$

An alternative approximate method of solution can be based on the f vs D* plot. This method avoids the use of trial and error so that it can be implemented without the use of a spread sheet and is correspondingly quite convenient.

Estimate the equivalent length of the fittings as follows:

Pipe entrance

$$\frac{L}{D} = 25$$

Elbows

$$\frac{L}{D} = 50$$

Tank entrance

$$\frac{L}{D} = 50$$

The pressure drop due to friction is obtained from the energy balance

$$-\frac{\Delta P_{\mathrm{f}}}{\rho_{\mathrm{f}}} + g\Delta z = 0$$

$$
\begin{aligned}
PGDTF &= -\frac{\Delta P_f}{L_{total}} = \frac{-g\rho_f \Delta z}{L_{pipe} + L_{fitting}} \\
&= \frac{9.81 \times 1000 \times 20}{45 + 25 \times 0.05 + 2 \times 50 \times 0.05 + 50 \times 0.05} \\
&= 3.65 \times 10^3 \ \mathrm{Pa/m} \\
D^{*3} &= \frac{D^3 \rho_f PGDTF}{2\mu_f^2} \\
&= \frac{(0.05)^3 1000 \times 3.65\,10^3}{2 \times 0.001^2} = 2.281 \times 10^3 \\
D^* &= 611.0
\end{aligned}
$$

From Figure 2.3 (use friction factor module in the FLUIDS toolbox as shown in Figure 2.16),

$$
\begin{aligned}
f &= 0.00506 \\
Re &= 2.123 \times 10^5 \\
\bar{V} &= \frac{Re\mu_f}{D\rho_f} \\
&= \frac{2.123 \times 10^5 \times 0.001}{0.05 \times 1000} \\
&= 4.25 \ \mathrm{m/s}
\end{aligned}
$$

2.6 Pumps

A pump will generally be required to move a fluid through any piping system. This is the device through which the required energy is introduced to the system. A wide variety of pumps is available to suit the many applications that arise in process engineering. The centrifugal pump is by far the most commonly used type of pump in the process industries and this is the only type of pump that will be considered here. The essential features of a centrifugal pump are shown in Figure 2.17.

2.6.1 Pump characteristic curves

The performance of a centrifugal pump is commonly expressed as the head that is generated by the pump at a specified flowrate or conversely as the flowrate

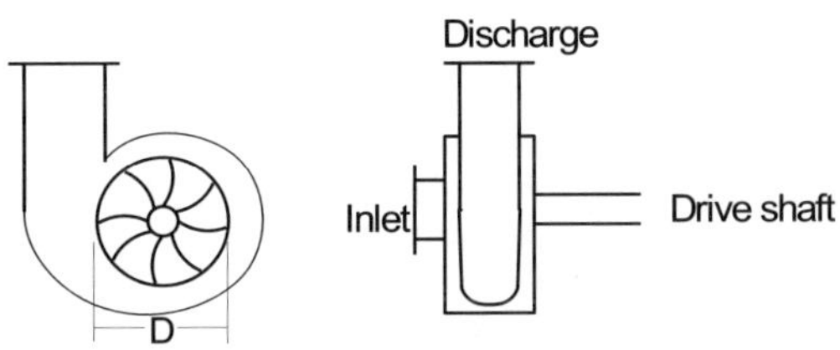

Figure 2.17 Essential features of a centrifugal pump

that can be delivered against a specified head. In practice a pump will rarely operate against a defined head or deliver a given flowrate. These variables are linked and are determined by the nature of the piping system through which the fluid must be delivered and by the speed at which the impeller rotates. A centrifugal pump generates a larger head the lower the flowrate delivered. The performance of a pump is usually determined in the laboratory by the manufacturer and is presented in the pump characteristic curve. The performance of a centrifugal pump is specified completely by the data that is presented on the pump characteristic curve. A typical pump characteristic curve for a commercial pump is shown in Figure 2.18. The characteristic curve shows primarily how the head developed by the pump varies as the discharge rate varies. In general the head decreases as the discharge rate increases. This decrease results from the hydrodynamic design of the pump and from the frictional dissipation in the pump chamber. The head that is developed at a particular throughput varies strongly with the speed of revolution of the pump and this is shown on the pump curve as a series of lines that represent the pump characteristic at each rotation speed. No pump is completely efficient in energy utilization and significant energy is lost between the mechanical drive on the shaft of the pump and the head that is developed to do useful flow work. The power transferred to the fluid is less than the work done by the impeller because of losses in the intake, impeller and pump chamber. Pump efficiency is defined as the ratio of useful hydraulic power delivered to the fluid to the power input at the drive shaft. The efficiency varies with the operating conditions and the iso-efficiency lines on the characteristic curve represent the efficiency at each

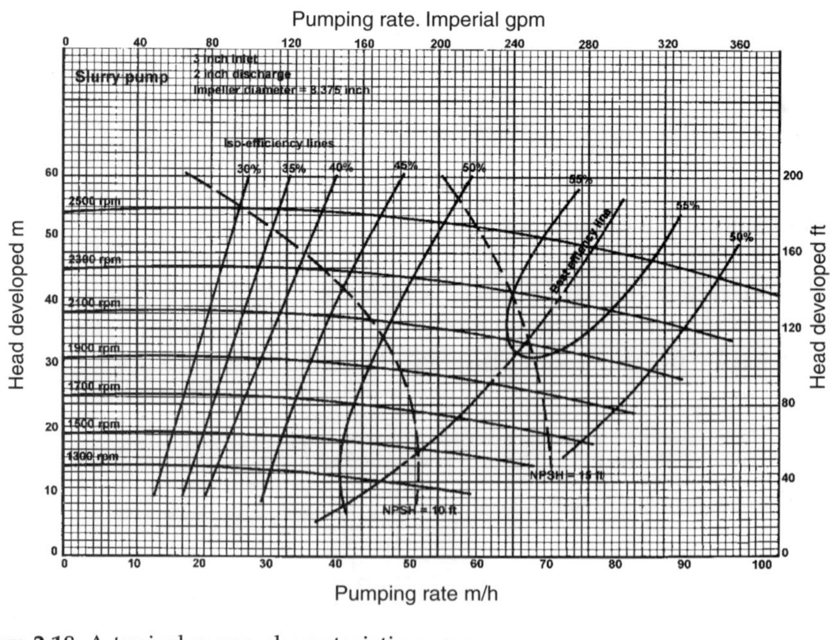

Figure 2.18 A typical pump characteristic curve

condition. This energy inefficiency must be taken into account when a motor is chosen to drive the pump for a specific application. At constant pump speed the efficiency increases as throughput increases and passes through a maximum before decreasing as the throughput becomes large. The efficiency also increases as the pump speed increases so the iso-efficiency contours are typically U-shaped curves as shown in Figure 2.18. The locus of the best efficiency points (BEP) is shown on the chart and the pump should be chosen so that the operating point is close to this line. The method for the establishment of the operating point is discussed later in this chapter.

2.6.2 The generalized pump characteristic curve

The centrifugal pump derives its pumping action from the centrifugal acceleration that is generated when the fluid rotates inside the pump chamber driven by the impeller. The centrifugal acceleration generates an increasing pressure from the center to the outer edge of the impeller. This pressure change is reduced by the frictional drag experienced by the fluid as it moves outward between the blades of the impeller shearing against the surface of the blades and against the inner surfaces of the pump casing.

A simple analysis provides a theoretical basis for the pump characteristic curve by analyzing the pressure generated by the rotating impeller. The method is based on the relationship between the torque that must be applied to the impeller and the change in the angular momentum of the fluid.

The torque applied to the impeller is equal to the rate of change of the angular momentum of the fluid as it moves through the pump from the inlet to the involute of the casing. Assume that the inlet fluid has no angular momentum then

$$\text{Torque} = T = u_\theta \rho_f Q \text{ Nm} \tag{2.45}$$

where Q is the volumetric pumping rate in m^3/s and u_θ is the tangential velocity of the fluid at the impeller tip. Equation 2.45 is referred to as Euler's turbomachinery equation.

The power required to drive the pump is obtained by multiplying the torque by the angular speed of the impeller. Thus the work that must be applied to the impeller per unit mass of fluid is

$$-W = \omega R u_\theta \text{ J/kg} \tag{2.46}$$

The energy balance over the pump can be described by equation 2.40.

$$W + F + \frac{\Delta P_{Pu}}{\rho_f} = 0 \tag{2.47}$$

ΔP_{Pu} is the increase in the fluid pressure across the pump.

$$\frac{\Delta P_{Pu}}{\rho_f} = \omega R u_\theta - F \text{ J/kg} \tag{2.48}$$

Because of the slip that is induced by the angle of the impeller blades, the fluid does not rotate at the same angular velocity as the impeller. This is

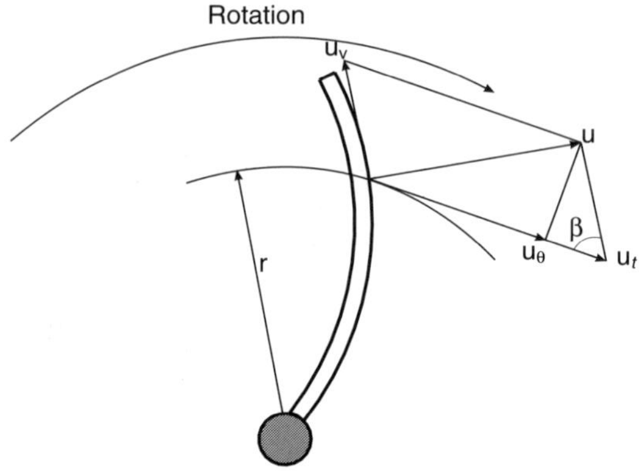

Figure 2.19 Schematic of a single impeller blade in a centrifugal pump

illustrated in Figure 2.19. At radius r the fluid velocity is represented by the vector u. The tangential velocity of the impeller at radius r is represented by the vector u_t. The tangential component of the fluid velocity is given by

$$u_\theta = u_t - u_v \cos\beta \tag{2.49}$$

where u_v is the slip velocity along the surface of the impeller vane, and using the angular speed of rotation ω of the impeller in place of u_t

$$u_\theta = \omega r - u_v \cos\beta \tag{2.50}$$

The increase in fluid pressure across the pump is calculated by substituting the value of u_θ at the impeller tip into equation 2.48

$$\Delta P_{Pu} = \rho_f \omega^2 R^2 - \rho_f \omega u_v R \cos\beta - 2\frac{\rho_f f u_v^2}{D_H} R \tag{2.51}$$

where D_H is the hydraulic mean diameter of the gap between the impeller and f the friction factor in the chamber. It is not possible to assign values to D_H or f and they must be estimated from experimental data.

The slip velocity u_v can be related to the volumetric flowrate through the pump chamber Q which is calculated as the outward radial component of the fluid velocity at the impeller tip.

$$Q = 2\pi R w u_v \sin\beta \tag{2.52}$$

where w is the width of the pump chamber.

The pressure that is developed across the pump chamber is usually designated in terms of the head of liquid that is generated. This is given by

$$\begin{aligned} h_{Pu} &= \frac{\Delta P_{Pu}}{g\rho_f} \\ &= \frac{\omega^2 R^2}{g} - \frac{\omega Q}{2\pi w g \tan\beta} - \frac{f Q^2}{2\pi^2 R w^2 \sin^2\beta D_H} \end{aligned} \tag{2.53}$$

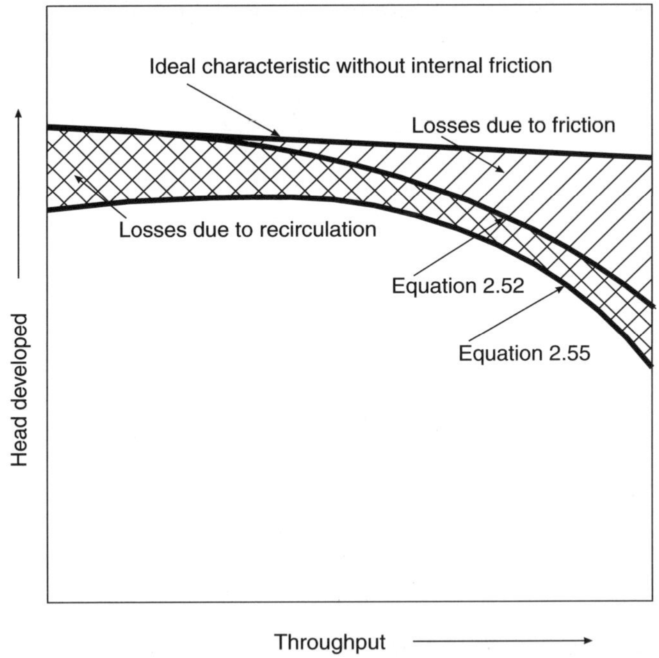

Figure 2.20 Schematic pump characteristic curve showing the contribution of various energy losses in the pump

The first two terms on the right hand side of equation 2.53 are referred to as the ideal pump characteristic because they show the expected performance in the absence of any frictional losses inside the pump chamber. The ideal characteristic is a straight line as shown in Figure 2.20. This line has negative slope for backward curved impellers ($\beta > 0$) and positive slope for forward curved impellers ($\beta < 0$)

Equation 2.53 shows that for a well-designed centrifugal pump, the dimensionless pump group $h_{P_u}g/\omega^2R^2$ should be 1.0 at zero pumping rate although in practice values less than 1.0 are found for real pumps because of shock and internal circulation losses that are not associated directly with the flow of fluid through the pump. These losses show up as an increase in the temperature of the fluid. Values for this group in the range 0.5 to 0.7 are common. This group is sometimes called the head coefficient or the dimensionless 'shut-off head' because it is based on the head that would be generated without any flow. Note from equation 2.53 that the head generated expressed as meters of pumped fluid is independent of the density of the fluid. This simple analysis also shows that the head generated should vary with the square of the pump speed. Although the head developed by the pump is independent of the fluid density, the pressure increase is proportional to the fluid density. This is the reason that a centrifugal pump needs priming because if it is filled with air it will not generate sufficient pressure to move the fluid.

The performance of a pump can be specified in terms of two dimensionless groups, the pump head number

$$N_{pu} = \frac{\Delta P_{Pu}}{N^2 D_{imp}^2 \rho_f} = \frac{h_{Pu} g}{N^2 D_{imp}^2} \tag{2.54}$$

and the pump flow number

$$N_Q = \frac{Q}{N D_{imp}^3} \tag{2.55}$$

Where D_{imp} is the impeller diameter and N the pump speed in revolutions per second.

A series of geometrically similar pumps will have w and D_H proportional to the impeller diameter and a generalized characteristic curve given by

$$N_{pu} = A - BN_Q - CN_Q^2 \tag{2.56}$$

$$A = \pi^2$$

$$B = \frac{1}{\gamma \tan \beta} \tag{2.57}$$

$$C = \frac{gf}{\pi^2 \gamma \delta \sin^2 \beta}$$

with $\gamma = w/D_{imp}$ and $\delta = D_H/D_{imp}$.

The parameters A, B and C are constants that depend only on the geometry of the pump but not on its size. The generalized characteristic curved is useful because pump manufacturers usually offer their pumps in geometrically similar series from small to large. These are called homologous series. A single generalized curve can be used to describe an entire series with a single set of parameters A, B and C. Although parameters A, B and C are given in terms of measurable geometrical properties of the pump in equation 2.57, they are more usefully considered to be empirical constants that must be determined from tests on pumps in the homologous series. The normal performance of a centrifugal pump operating without cavitation can be defined by specifying values for the two dimensionless groups N_Q and N_{pu}. Equation 2.56 describes the performance of the entire pump series. The parameter C is always positive while B can be positive or negative. Once values of A, B and C have been determined for a particular pump, the actual pump characteristic curve can be easily constructed from the generalized curve. The characteristic curve for any pump in the series can be generated from the generalized curve by substitution of the appropriate value of impeller diameter D_{imp} and rotation speed into equations 2.54, 2.55 and 2.56.

Values of A, B and C can be determined experimentally by measuring the head developed by the pump at different delivery rates and rotation speeds.

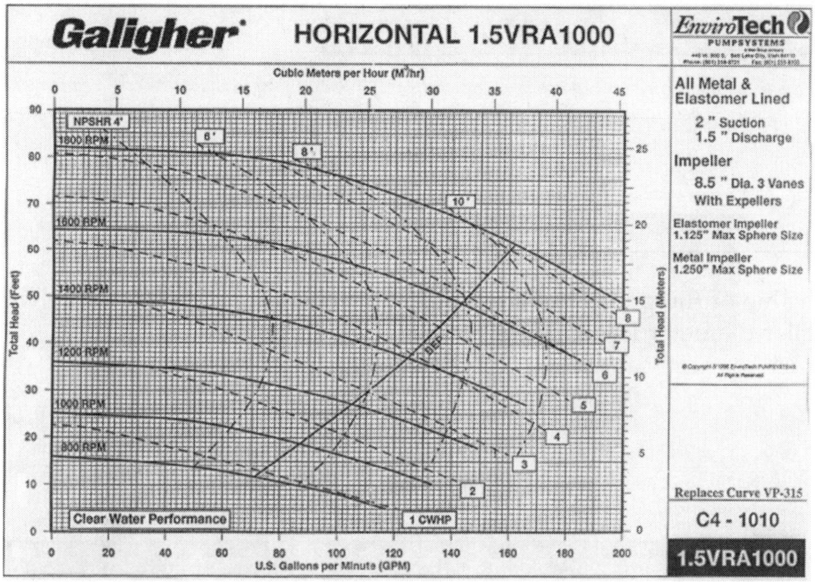

Figure 2.21 Characteristic curve for a Galigher 1.5 VRA 1000 pump. Published by courtesy of Weir Slurry Group, Inc

Pump manufacturers always supply this information for their pumps in graphical form in which the head developed is plotted against the delivery rate at different rotation speeds. Typical examples of manufacturers curves are shown in Figures 2.18 and 2.21. The parameters can also be obtained by fitting equation 2.56 to the measured or manufacturer's pump characteristic curve or, if that is not available, to the curve of a different size pump in the same series. Equation 2.56 is especially convenient for computer solution because of its convenient form. The generalized curve makes it easy to interpolate accurately to pump impeller speeds that are not included specifically on the manufacturer's curve. However, the generalized curve should not be used as a substitute for the actual characteristic curve determined from tests by the pump manufacturer. Note that the parameters A, B and C do not remain constant when impellers of varying diameter are fitted into the same pump casing because the clearance geometry between impeller and casing does not remain constant.

Illustrative example 2.7 Generalized pump characteristic curve
The following data can be read from the pump characteristic curve shown in Figure 2.21. Show that these data are consistent with the generalized pump characteristic curve.

Connections:	2″ suction, 1.5″ discharge.
Size:	8.5″ diameter 3-vane impeller.

Pump speed revs/min	Shut off head ft of water
800	16.1
1000	25.1
1200	36.1
1400	49.3
1600	64.3
1800	82.0

The value of the pump number is calculated at the pump speeds given. The impeller diameter is 8.5 inches or 0.2159 m and at 800 revs/min.

$$N_{pu} = \frac{h_{pu}g}{N^2 D_{imp}^2} = \frac{4.91 \times 9.81}{13.33^2 \times 0.2159^2} = 5.81$$

$$\text{Head coefficient} = \frac{N_{pu}}{\pi^2} = 0.589$$

Pump speed		Shut off head		Pump head number N_{Pu}	Head coefficient
rpm	revs/s	ft	m		
800	13.33	16.1	4.91	5.81	0.589
1000	16.67	25.1	7.65	5.80	0.588
1200	20.00	36.1	11.00	5.79	0.587
1400	23.33	49.3	15.03	5.81	0.589
1600	26.67	64.3	19.60	5.80	0.588
1800	30.00	82.0	24.69	5.84	0.592

The pump number N_{pu} at zero flow is very nearly constant over the whole range of impeller speeds as required by the simple theory.

Illustrative example 2.8

The line for clear water horsepower (CWHP)=4 passes through the coordinates $Q = 140$ US gal/min and $h_{pu} = 33.3$ ft of water on Figure 2.21. Calculate the pump efficiency at this condition.

The data read from the curve should be converted to the SI before any calculations are started using the SI units converter in the FLUIDS toolbox.

$$\text{Power} = P = 4\text{HP} = 2.983 \times 10^3\ \text{W}$$

$$Q = 8.8325 \times 10^{-3}\ \text{m}^3/\text{s}$$

$$h_{Pu} = 10.15\ \text{m}$$

$$\varepsilon = \frac{g\rho_f h_{Pu} Q}{P} = \frac{9.81 \times 1000 \times 10.15 \times 8.8325 \times 10^{-3}}{2.983 \times 10^3}$$

$$= 0.295$$

2.6.3 Derating of pumps when handling slurries

The simple theory for the pump characteristic curve that is described in the previous section shows that the head developed by a centrifugal pump, measured as head of the fluid being pumped, is independent of the fluid density. This is convenient in practice because the same characteristic curve can be used for fluids of different densities. However, when the pump must transport a slurry, the presence of the solid particles has a significant effect on the performance of the pump. The performance of the pump is derated to account for this. As the concentration of solids in the slurry increases the head generated by the pump decreases because of the greater frictional losses that occur in the pump casing as the slurry moves through. The reduction in performance is described quantitatively in terms of the head reduction ratio relative to the head that would be produced without solids in the carrier fluid.

This reduction ratio depends primarily on the volume fraction of the solids in the slurry. Data showing the head reduction ratio as a function of volumetric concentration are presented in Figure 2.22. The slurries used to generate the data shown in Figure 2.22 were made from beach sand and river sand. The sand was predominantly quartz having density 2670 kg/m^3. The river sand was considerably coarser than the beach sand. The median size of the beach sand was 295μm and that of the river sand was 1290μm. The head reduction ratio is seen to decrease linearly over the concentration range from

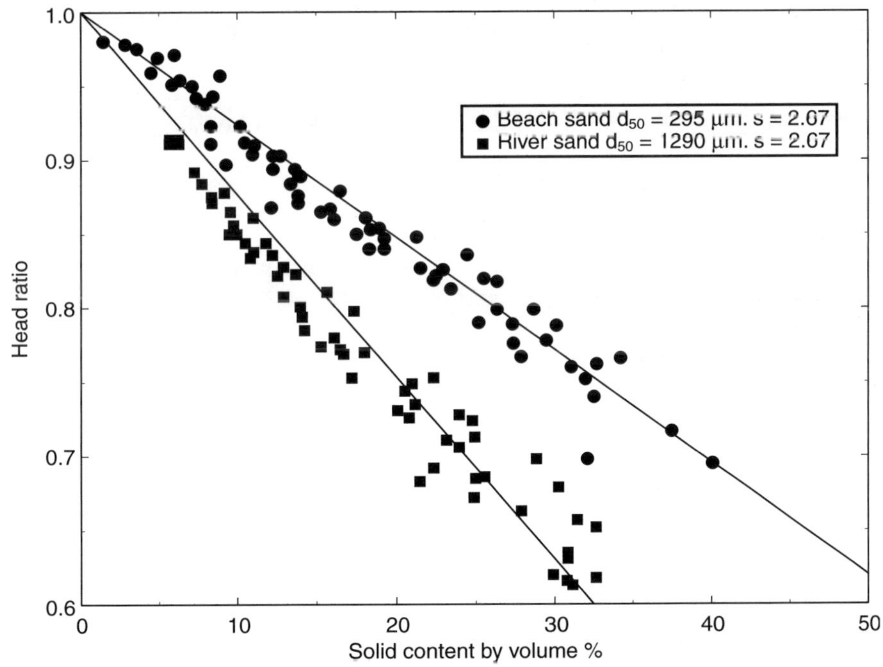

Figure 2.22 Head reduction ratios for pumping of slurries in centrifugal pumps. Data from Cave (1976)

0 to 40 per cent solids by volume. A useful model for the derate phenomenon is a simple linear relationship between the head ratio and the volume fraction of solids.

$$H_r = 1 - KC_v \tag{2.58}$$

where H_r is the head reduction ratio and C_v is the volume fraction of solids in the slurry. The slope of the line through the data in Figure 2.22 varies with a number of operating variables. After the volume fraction of the solid, the next most significant variable is the size of the particles in the slurry. Coarser particles show the greatest effect and this can be clearly seen in Figure 2.22. The variation with particle size is approximately logarithmic and K can be evaluated from

$$K = 0.30 \ln\left(\frac{d_{50}}{d'}\right) \quad \text{for} \quad d_{50} > d' \tag{2.59}$$

where d_{50} is the median size of the particle population in the slurry. This equation implies that the head ratio has a value of 1.0 at median particle size of d' and below so that slurries made from very small particles behave in the pump chamber as uniform liquids and the pump performance is not derated. In older pumps d' must be as low as about 25μm to reduce derating to negligible proportions but in newer pumps with specially designed impellers d' can be as high as 105μm.

Other factors that affect the value of K are the density of the solid that makes up the slurry and the flowrate through the pump. Experiments with ilmenite and heavy mineral sand indicate that K varies in proportion to the factor $s - 1$ where s is the specific gravity of the solid. Thus K can be calculated from

$$K = 0.18(s - 1) \ln\left(\frac{d_{50}}{d'}\right) \tag{2.60}$$

The variation of the head ratio with flowrate and pump speed is not very pronounced and is usually neglected.

2.6.4 Pump efficiency

A centrifugal pump does not convert to useful flow energy all of the power that is supplied to its drive shaft by the electric motor. Considerable energy is dissipated in the bearing which must be sufficiently tight to prevent leakage – either inward or outward depending on whether the absolute pressure inside the pump is below or above the atmospheric pressure respectively. Internal fluid friction also accounts for sizeable amounts of energy dissipation. Energy losses range from about 10 per cent of the drive energy to as much as 80 per cent. The energy efficiency must be taken into account when evaluating the performance of any pump.

The energy efficiency of a pump can be represented in different ways on the pump characteristic curve. Many pump manufacturers determine this by experiment and plot iso-efficiency contours on the pump characteristic curve as shown in Figure 2.18.

An additional dimensionless group, the pump specific speed

$$N_{\mathrm{Sp}} = \frac{N_Q^{1/2}}{N_{\mathrm{pu}}^{3/4}} = \frac{NQ_{\mathrm{BEP}}^{1/2}}{(h_{\mathrm{Pu\,BEP}}g)^{3/4}} \tag{2.61}$$

is used to specify the approximate pumping capacity of a pump relative to the head developed. It is customary to specify the specific speed at the best efficiency point (BEP). The reader should note that many authors omit the gravitational acceleration from the denominator of equation 2.61 and often use inconsistent units for the other terms. This destroys the nondimensional character of the specific speed and this practice is not used in this book. The power required to drive the pump is given by

$$P = \frac{W\rho_f Q}{\varepsilon} = \frac{\Delta P_{\mathrm{pu}} Q}{\varepsilon} = \frac{g\rho_f h_{\mathrm{Pu}} Q}{\varepsilon} \text{ W} \tag{2.62}$$

where ε is the pump efficiency.

It is convenient to incorporate the pump efficiency on the pump characteristic curve since this will ultimately determine the power required for the pumping application. Three different methods of displaying this data are commonly used and different manufacturers usually choose one of these. The iso-efficiency contours as shown in Figure 2.18 allow the efficiency to be estimated approximately at any point on the graph. Other manufacturers prefer to draw contours of constant required input energy as shown by the lines labeled as CWHP (clear water horse power) in Figure 2.21. Example 2.8 illustrates the relationship between these two methods. The third method shows efficiency plotted against the pump throughput at various pump speeds. These methods provide the same information and are equivalent.

For most pumps the efficiency increases with flowrate at a constant pump speed and passes through a maximum before decreasing quite sharply. The combination of Q and h_{Pu} that gives the maximum efficiency at any particular pump speed is called the best efficiency point usually abbreviated to BEP. The locus of the points of maximum efficiency is usually shown on the pump characteristic curve and it is usually recommended that the operating point for the pump should be located on this locus. The BEP locus is shown in Figures 2.18 and 2.21.

The BEP locus can be determined from the specific speed of the pump. The specific speed may be taken as independent of the rotation speed of the impeller over the normal working range of the pump. Once N_{Sp} is specified for the pump the locus of BEP can be calculated and plotted on the pump characteristic curve by solving equations 2.56 and 2.61. This requires the simultaneous solution of two non-linear equations and is done numerically. The solution is implemented automatically in the FLUIDS toolbox when a pump characteristic curve is generated. Dimensionless specific speeds for radial flow centrifugal pumps vary over the range 0.03 to about 0.15. Axial flow pumps have higher specific speeds because they deliver higher pumping rates at equivalent heads.

An approximate generalized method for specifying the pump efficiency is based on the postulate that the efficiency at any operating condition relative to the efficiency at the BEP is a generalized function of the ratio of the pump flow number N_Q at the operating point to the pump flow number at BEP. Experimental data reveal that this function has a shallow maximum at $N_{Q\ \mathrm{BEP}}$ and a useful approximate relationship is

$$\frac{\varepsilon}{\varepsilon_{\mathrm{BEP}}} = 2\frac{N_Q}{N_{Q\,\mathrm{BEP}}}\left(1 - \frac{N_Q}{2N_{Q\,\mathrm{BEP}}}\right) \tag{2.63}$$

In addition the BEP efficiency is known to decrease as the pump speed decreases and a simple linear relationship is assumed here with a reduction of 1 per cent in efficiency for every 120 rpm reduction in pump speed. These simple relationships are used in the FLUIDS toolbox to plot iso-efficiency lines on the pump characteristic curve. The use of the toolbox to generate a pump characteristic curve is illustrated in Figures 2.23 and 2.24.

The pump efficiency is decreased by the presence of solids in much the same way as the developed head. A simple model that can be used is to assume that $\varepsilon_r = H_r$. However, this should be regarded only as a rough first approximation.

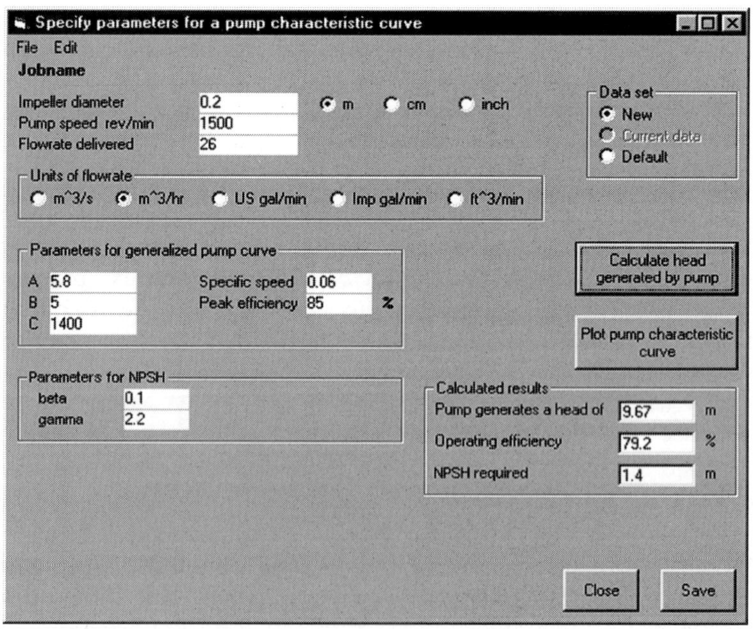

Figure 2.23 Data input to generate a pump characteristic curve using the FLUIDS toolbox

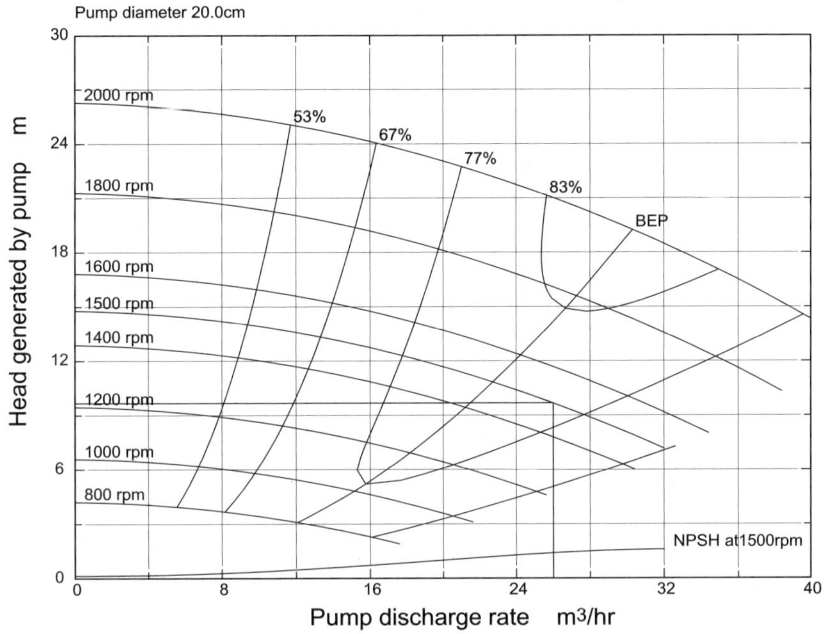

Figure 2.24 Pump curve generated by the FLUIDS toolbox using data shown in Figure 2.23

2.6.5 Net positive suction head

The pumping action of the impeller creates a reduction in pressure in the interior of the pump. This reduction in pressure creates a potential problem that must be considered when evaluating the duty of the pump. The lowest pressure in the pump chamber is usually considerably below the pressure at the inlet flange and it is important that the lowest pressure anywhere in the pump chamber should never fall below the vapor pressure of the liquid that is being pumped. If the pressure does fall below the vapor pressure of the liquid, the liquid will spontaneously vaporize forming vapor bubbles. As soon as any bubble moves into a region of higher pressure in the pump it will collapse rapidly and the liquid that rushes into the collapsing bubble will strike a violent blow to the impeller or other surface in the pump. This phenomenon is known as cavitation and it can cause considerable damage to the impeller and seriously reduce the pumping capacity of the pump. The presence of vapor bubbles in the pump chamber also decreases the effective density of the fluid that is being pumped. This will lead to a significant reduction in the outlet pressure generated by the pump as can be seen from equation 2.53.

In order to prevent cavitation it is necessary to ensure that the pressure at the inlet flange exceeds the vapor pressure of the liquid by at least the pressure drop that will be generated between the inlet and the point of lowest pressure in the pump chamber. This excess pressure is called the net positive suction head that must be provided at the inlet flange to prevent cavitation.

In practice this means that the supply reservoir will normally have to be above the level of the pump when the pump is fed by gravity. Normally the level of the liquid in the feed reservoir will need to be elevated above the feed point by more than the net positive suction head because there will certainly be some frictional dissipation in the line from the reservoir to the pump. Remember that it is the absolute pressure (not the gauge pressure) at the inlet that must exceed the sum of the vapor pressure and the NPSH.

$$P_{\text{in}} > \text{Vapor pressure} + \text{NPSH} \tag{2.64}$$

The NPSH of a pump varies with the volumetric flow through the pump – larger flowrates generating larger pressure drops from the inlet to the interior and therefore requiring larger suction heads. The NPSH is usually shown as a series of lines on the pump characteristic curve. Four lines are shown in Figure 2.21 for NPSH of 4, 6, 8 and 10 feet of liquid. The shape of the NPSH lines vary markedly from pump to pump since they depend strongly on the mechanical design of the pump chamber and the impeller.

The NPSH that is required by a pump depends to a considerable extent on the mechanical design of the impeller and the pump chamber. NPSH values that are shown by various pump manufacturers on their published pump characteristic curves vary quite markedly and no general theoretical model for the required NPSH has been developed from simple assumptions such as those that led to the generalized pump characteristic curve. A simple correlating principle that has been used with some success is to assume that the parameter $\sigma = \text{NPSH}/\text{TDH}$ is a function of the ratio $Q' = Q/Q_{\text{BEP}}$ and is independent of pump rotation speed. TDH is the total delivered head at the throughput Q. A useful empirical expression that matches a number of manufacturers' data is

$$\sigma = 0.01 + \beta Q'^{\gamma} \tag{2.65}$$

$\beta + 0.01$ is the required NPSH at Q_{BEP} expressed as a fraction of h_{BEB}. Values of β in the range 0.05–0.15 have been found for commercial pumps. γ is a parameter that defines the rate of increase of NPSH with throughput. Values for γ in the range 1.3–2.5 will probably cover the requirements for common centrifugal pumps. Equation 2.65 implies that a NPSH of 1 per cent of the shut off head must be available at no flow to prevent cavitation.

Equation 2.65 does not imply that NSPH is independent of the rotation speed because Q_{BEP} varies with rotation speed as can be seen from the BEP locus on Figures 2.18, 2.24 and 2.21. The NPSH calculated from equation 2.65 is shown on the pump characteristic curves that are generated by the FLUIDS toolbox. An example is shown in Figure 2.24.

2.6.6 Pump performance and the system curve

The pump characteristic curve can be used effectively to select a suitable pump for the duty that is required and it can also be used to establish the capacity of a pipeline system as a function of the speed of a pump that already exists in the pipeline. The physical details and frictional characteristics of the pipeline and all fittings and included equipment must be known. The energy balance for the system is given by equation 2.66

$$-W = \sum_i F_i + v\Delta P + g\Delta z + \frac{1}{2}\Delta\bar{V}^2 \tag{2.66}$$

where F_i is the frictional dissipation in segment i of the pipeline system.

The work required by the pump can be written in terms of the pressure or head developed by the pump

$$h_{\text{Pu}} = \frac{-W}{g} = \frac{\sum F_i}{g} + \frac{\Delta P}{\rho_f g} + \Delta z + \frac{\Delta\bar{V}^2}{2g} \tag{2.67}$$

The head developed by the pump must match the head required to drive the system as given by the solution to equation 2.67. This is illustrated in Figure 2.25 where the right hand side of equation 2.67 is plotted against Q on the pump characteristic curve graph. The velocity in the pipe in equation 2.67 is related to the flowrate by

$$\bar{V} = \frac{4Q}{\pi D^2} \tag{2.68}$$

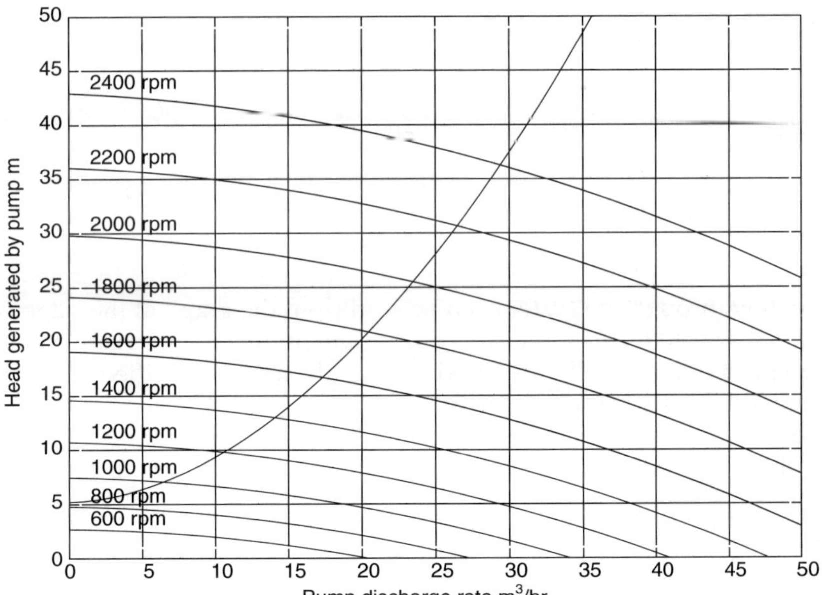

Figure 2.25 Pump characteristic curves with a system curve. Pump has impeller diameter 8.5″ and generalized characteristic curve is $N_{\text{pu}} = 5.846 + 11.23N_Q - 1570N_Q^2$. The system consists of 100 m 5 cm ID pipe with roughness = 0.02 mm. The static head in the system is 5.2 m.

The frictional dissipation in each segment of straight pipe is given by equation 2.7.

$$F_i = 2f\bar{V}_i^2 \frac{L_i}{d_i} \tag{2.69}$$

The frictional dissipation due to each fitting can be calculated using the loss coefficient (equation 2.44)

$$F_i = \frac{K_{fi}\bar{V}_i^2}{2} \tag{2.70}$$

The operating point is located at the intersection of the system curve and the appropriate pump characteristic curve.

The method is illustrated for a simple specific case in which the pump is assumed to have the generalized characteristic curve

$$N_{pu} = 5.846 + 11.23N_Q - 1570N_Q^2 \tag{2.71}$$

The system consists of 100 m of 5 cm ID pipe with roughness = 0.02 mm. The static head for the system is 5.2 m The system curve is calculated from the equation

$$h_{Pu} = \frac{2f\bar{V}^2 L}{gD} + 5.2 \tag{2.72}$$

with

$$\bar{V} = \frac{4Q}{3600\pi D^2} \tag{2.73}$$

In practice the system usually consists of many segments of pipe of varying length and varying diameter and several fittings. The head loss must then be calculated for each pipe segment and fitting and the total system curve plotted.

The flow through the system can be read from the graph at the intersection between the system curve and the appropriate pump characteristic curve. The pump speed that would be required to provide any prespecified flowrate can be easily established from the diagram.

The system curve can be superimposed on the pump characteristic curve using the FLUIDS software toolbox (Professional version only).

---

llustrative example 2.9

6.5 liters/second of water must be transferred between two reservoirs both of which are open to the atmosphere. The water surfaces in the reservoirs differ by 30 m in elevation. The pipe is 2 km straight run of 5.25 cm internal diameter new commercial steel (roughness 0.045 mm). Allow for the frictional dissipation at the exit from the first reservoir and at the entrance to the second reservoir. Assume that there is 5 m of pipe between the upstream reservoir and the first pump. What is the power requirement for this duty? Select suitable pumps for this duty.

Internal diameter of pipe = 5.25 cm

Internal cross-sectional area $= 0.25\pi(5.25 \times 10^{-2})^2 = 2.165 \times 10^{-3}\,\text{m}^2$

Velocity in the pipe $= \bar{V} = \dfrac{6.5 \times 10^{-3}}{2.165 \times 10^{-3}} = 3.00\,\text{m/s}.$

$$Re = \frac{d\bar{V}\rho_\text{f}}{\mu_\text{f}} = \frac{0.0525 \times 3.00 \times 1000}{0.001}$$

$$= 1.576 \times 10^5$$

Roughness = 0.045 mm

Relative roughness $= \dfrac{0.045}{52.5} = 0.000857$

Friction factor $f = 0.005028$ from FLUIDS toolbox.

$$F = 2f\frac{\bar{V}^2 L}{D}$$

$$= \frac{2 \times 0.005028 \times 3.00^2 \times 2 \times 10^3}{0.0525}$$

$$= 3.448 \times 10^3$$

$PGDTF = 3.448 \times 10^3/2000 = 1.724\,\text{kPa/m}$

It is useful to have this expressed in m of head per meter of pipeline length.

$$PGDTF = \frac{1724}{9.81 \times 1000} = 0.176\,\text{m head/m pipe length}$$

For the pipe exit at reservoir 1,

$$F = K_\text{f}\frac{\bar{V}^2}{2} = \frac{1.0 \times 3.0^2}{2} = 4.5 \text{ from Table 2.5}$$

For the pipe entrance to reservoir 2,

$$F = K_\text{f}\frac{\bar{V}^2}{2} = \left(\frac{K_1}{Re} + K_\infty\right)\frac{\bar{V}^2}{2}$$

$$= \left(\frac{160}{1.576 \times 10^5} + 1.0\right)\frac{3^2}{2} = 4.5$$

Overall energy balance

Use reservoir levels as reference planes.

$$v = \frac{1}{\rho_\text{f}}$$

$$W + F + \frac{\Delta P^{\nearrow 0}}{1000} + g\,30 + \frac{1}{2}\Delta\bar{V}^{2\nearrow 0} = 0$$

$$W = -3.448 \times 10^3 - 4.5 - 4.5 - 9.81 \times 30 = -3.751 \times 10^3\,\text{J/kg}$$

To convert this to pressure developed by the pump, multiply by ρ

$$\Delta P_{\text{Pu}} = -W \times 1000 = 3.751\,MPa$$

$$h_{\text{Pu}} = \frac{\Delta P_{\text{Pu}}}{g\rho} = 382.4\,m\ of\ water$$

To convert to kW, multiply by mass flowrate

$$\begin{aligned}\text{Power} &= 3.751 \times 10^3 \times 6.5 \times 10^{-3} \times 1000 \\ &= 24.3 \times 10^3\,W = 24.3\,kW \\ &= \frac{24.3}{0.740} = 32.6\,HP\end{aligned}$$

Consult a pump selection chart to find that a 3 × 2 pump will probably be required for this duty but that the required head can not be achieved in a single pump. This means that several pumps must be inserted along the length of the pipeline. Using the characteristic curve for a 3 × 2 pump (Figure 2.18), the design for this system can be completed. This pump, running at 2400 rpm, will generate a head of 50 m at $Q = 23.4\,m^3/hr$ and will require a NPSH of 2.75 m. A simple graphical technique can be used to locate the position of each pump in the pipeline as shown in the figure 2.26.

$$\text{Hydraulic gradient} = \text{slope} = -0.176\,\text{m/m}$$

Start at the upstream end. The first pump generates a head of 50 m and the head is reduced at the rate of 0.176 m per m of pipeline length as shown by the sloping line. The second pump is located at the intersection of this sloping line and the surface grade that is represented by the line that slopes upward at 30/2000 m/m pipe length. This ensures that the inlet pressure in the second pump is the same as that in the first pump in spite of the increased elevation of the second pump. The succeeding pumps can be located in exactly the same way as shown in the diagram. Eight pumps will be necessary to provide the required total head.

Next it is necessary to determine the elevation of the first pump to ensure that the required NPSH for the pump is available?

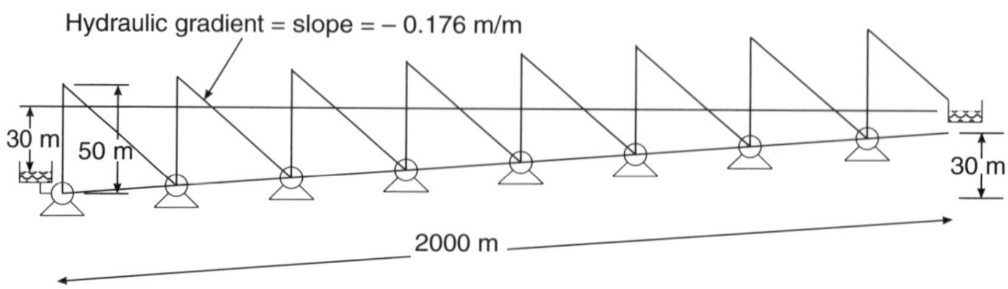

Figure 2.26 Graphical construction to locate pumps along the pipeline

NPSH for the pump which can be read from the characteristic curve for the pump. Using the characteristic curve for the 3 × 2 pump, the required NPSH is 2.75 m of water which is equivalent to 27.0 kPa. The vapor pressure of water at 25° is 3.17 kPa.

The pressure required at the suction side of the pump is 27.0 + 3.17 kPa. The diameter of the pump inlet pipe is 5.25 × 3/2 = 7.85 cm and the velocity in the suction pipe is $3.00 \times 2^2/3^2 = 1.33$ m/s.

$Re = 0.0785 \times 1.33 \times 1000/0.001 = 1.044 \times 10^5$

The friction factor f = 0.00432

Energy balance between upstream reservoir surface (A) and the pump inlet (B)

$$P_A = 1.013 \times 10^2 \text{ kPa},$$
$$P_B = 30.17 \text{ kPa},$$
$$\Delta P = -71.13 \text{ kPa}$$

$$W + F + v\Delta P + g\Delta z + \frac{1}{2}\Delta \bar{V}^2 = 0$$

$$F = K_f \frac{\bar{V}^2}{2} + 2f\bar{V}^2 \frac{L}{D}$$
$$= \left(\frac{160}{1}.044 \times 10^5 + 1.0\right) \frac{1.33^2}{2} + 2 \times 0.00432 \times 1.33^2 \frac{5}{0.0785}$$
$$= 0.886 + 0.973 = 1.859$$

$$1.859 + \frac{-71.13 \times 10^3}{1000} + 9.81\Delta z + 1.33^2/2 = 0$$
$$\Delta z = 9.97 \text{ m}$$

The pump inlet can be as much as 9.97 m above the level of the water in the upstream reservoir.

A better design would use a larger diameter pipe to decrease the hydraulic gradient and thereby reduce the number of pumps that would be required.

2.7 Symbols used in this chapter

A Constant in generalized pump characteristic curve.
B Constant in generalized pump characteristic curve.
BEP Best efficiency point.
C Constant in generalized pump characteristic curve.
D Pipe diameter, m.
D_H Hydraulic mean diameter, m.
D_{imp} Impeller diameter, m.

D^*	Dimensionless pipe diameter.
e	Roughness of the surface of the pipe wall, m.
f	Friction factor.
F	Energy dissipated by friction per unit mass of fluid, J/kg.
g	Gravitation constant, m/s^2.
h_f	Head loss due to friction, m of fluid.
h_{Pu}	Head developed by pump, m of pumped fluid.
K_f	loss coefficient for pipe fitting.
L	Length of pipe, m.
N	Pump speed, revs/s.
N_{pu}	Dimensionless pump head number.
N_Q	Dimensionless pump flow number.
N_{Sp}	Dimensionless specific speed of a centrifugal pump.
N_{vh}	Number of velocity heads.
$NPSH$	Net positive suction head, m. of fluid.
P	Power drawn by pump, W.
$PDGTF$	Pressure gradient due to friction, Pa/m.
Q	Volumetric flowrate, m^3/s.
Q^*	Dimensionless flowrate.
Re	Reynolds number.
T	Torque, Nm.
$\bar{V}$	Average velocity in pipe, m/s.
V^*	Dimensionless average velocity.
v	Specific volume of fluid, m^3/kg.
β	Diameter ratio for pipe fittings.
β	Impeller vane angle in centrifugal pump.
ΔP_f	Pressure drop due to friction, Pa.
ε	Pump efficiency.
μ_f	Viscosity of fluid, Pa s.
ρ_f	Density of fluid, kg/m^3.
τ_w	Shear stress at the pipe wall, N/m^2.

2.8 Practice problems

1. Water is transferred between two reservoirs that are situated 3 km apart. The flow is by gravity and the difference in water level between the two reservoirs is 107.8 m. The pipeline diameter is 15 cm and the surface roughness is 0.085 mm.

   (a) Calculate the rate at which water is transferred between the reservoirs.

   (b) What pipe diameter should be used to increase the flow to 215 m^3/hr.

   $\mu_w = 0.001$ kg/ms, $\rho_w = 1000$ kg/m^3, g $= 9.81$ m/s^2.

2. An experiment is conducted to estimate the inside surface roughness of a pipe so that a pipeline can be selected to convey a water supply.

In the experiment, water flowed at 45 liters/s through a horizontal 15 cm ID pipe. The pressure drop was measured to be 15.4 kPa over a length of 30 m.

(a) Calculate the Reynolds number.
(b) Calculate the roughness of the pipe wall.
(c) Calculate the pressure drop that would be required to move 45 liters/s through 15 km of the pipe. Express the result in MPa and equivalent head of water.
(d) How many velocity heads are lost due to friction in the long pipeline?

Density of water is 1000 kg/m^3 and viscosity of water is 0.001 kg/ms.

3. The viscosity of a liquid can be estimated by measuring the rate of flow through a smooth tube of known cross-section. An open vessel drains through a horizontal tube 20 cm long and 2 mm internal diameter. When the depth of the liquid in the vessel is 6 cm, the fluid was found to discharge at 1.16 cm^3/s. The density of the fluid is 0.97 g/cm^3. Calculate the fluid viscosity.
4. Calculate the power required to drive a pump that transfers 100 per cent acetic acid (specific gravity = 1.0, viscosity = 1.155×10^{-3} kg/ms) at a rate of 3 liters/s from a vessel at ground level into a storage tank that is 7 m above ground level. The pipe is 4 cm ID stainless steel (absolute roughness = 0.002 mm). The vapor space in the lower vessel is at atmospheric pressure but the upper tank is pressurized at 10 MPa. The liquid is discharged into the vapor space in the upper tank. The liquid level is 0.75 m above ground level in the lower vessel and the discharge is 35 cm above the bottom of the upper tank. The pipe is 10.45 m long and pressure loss in fittings is equivalent to 5 elbows ($K_1 = 800$, $K_\infty = 0.40$). Atmospheric pressure = 98 kPa. When the pump is running, the liquid level in the lower vessels drops at a rate of 2.2 mm/s. The efficiency of the pump is 68 per cent.
5. Find the drop in pressure due to friction in a pipe 300 m long and 10 cms ID when water is flowing at the rate of 200 kg/s. The surface roughness is equal to 0.02 mm. Viscosity of water is 10^{-3} kg/s and density is 10^3 kg/m^3.
6. 5000 kg/hr of water are to be pumped through a smooth steel pipe 25 mm diameter and 100 m long to a tank. The water level in the tank is 10 m higher than the level in the reservoir from which the water is pumped. Calculate the power required by the pump motor. The pump runs at 35 per cent efficiency. Both the tank and the reservoir are at atmospheric pressure and differences in kinetic energy in tank and reservoir can be neglected.
7. The following values were read from a manufacturer's pump characteristic curve. Calculate the power required to drive a pump at this operating point.
   $Q = 822$ liters/s
   $h_{Pu} = 80.3$ m
   $\varepsilon = 79\%$.
8. A liquid is pumped 2 km from an open reservoir to an open storage reservoir through a 15 cm internal diameter pipe at a rate of 50 kg/s. What

is the pressure gradient along the pipe, and what power must be supplied to the pump if it has an efficiency of 50 per cent? The static head in the pipe is 10 m.

After some time the pump impeller becomes eroded and the gauge pressure at its delivery falls to one half of the value when the pump was new. By how much is the flow rate reduced? The pump discharge is 15 m below the level of the liquid in the reservoir.
Specific gravity of the liquid = 0.705
Viscosity of the liquid = 0.5×10^{-3} kg/ms
Roughness of the pipe surface = 0.04 mm.

9. Calculate the maximum flowrate of water through a 50 m length of 5 cm diameter mild steel pipe. The water is pumped by a pump running at 1800 rpm. The characteristic curve is given in Figure 2.21. The discharge is 8m above the suction and the line contains one fully open gate valve. What power does the pump motor draw when the valve is throttled down to give a flow of 5 m^3/hr through the line? Suction and discharge are at atmospheric pressure.
10. A series of centrifugal pumps has a generalized characteristic curve given by

$$N_{\text{pu}} = 1.77 - 1.81 \times 10^4 N_Q^2$$

What speed would be required for a pump with a 35 cm diameter impeller in this series to deliver 0.0035 m^3/s of water into a piping system that is equivalent to 50 m of 25 mm ID pipe with a static head of 30 m? Suction and discharge are both at atmospheric pressure and changes in kinetic energy can be neglected. The friction factor can be calculated from a Blasius equation

$$f = 0.077\, Re^{-0.25}$$

Viscosity of water = 0.001 kg/ms
Density = 1000 kg/m^3.

11. What head would be generated by a pump with characteristic curve given in Figure 2.18 when a slurry of river sand is pumped at 65 m^3/h. The pump speed is 2100 rpm and the solid content of the slurry is 20 per cent by volume.
12. The diagram shows a bridge-mounted thickener underflow pump (see Figure 2.27). The three pulp zones in the thickener have different densities which are identified in the diagram. At the specified pumping rate, the NPSH required by the pump is 1.5 m water. Can the pump be used in the position shown? Calculate the power required by the pump if the pump discharge pressure is 600 kPa and the pump is 30 per cent efficient. The solid flowrate in the underflow is 10 kg/s. The friction factor in the suction line is 0.02. The specific gravity of the solid is 2.67. Diameter of the suction line is 10 cm. Vapor pressure of water is 30 kPa. Atmospheric pressure is 100 kPa.

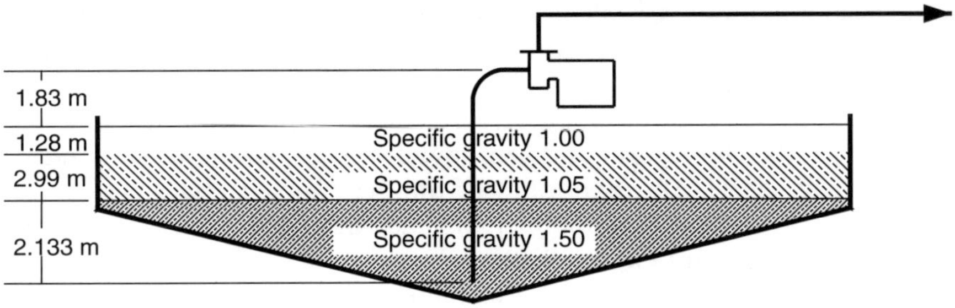

Figure 2.27 Cross-section of a thickener

13. A pump delivers water into a pipeline consisting of 100 m of smooth 25.4 cm ID pipe. The static head is 25 m. The pump runs at 1000 rpm and delivers 900 m^3/hr at a head of 45 m of water. The required flowrate is obtained by throttling a control valve in the delivery pipe. Calculate the power required to drive the pump.

    If the control system is changed to a variable speed drive on the pump, what speed must the pump run at to deliver the required flowrate if the control valve is removed from the line? What is the saving in power compared to that used with the throttle control valve?
    Density of water = 1000 kg/m^3.
    Viscosity of water = 0.001 kg/ms.
14. Water at 20 °C is to be pumped at a rate of 300 gpm from an open well in which the water level is 100 ft below ground level into a storage tank that is 80 ft above ground. The piping system contains 700 ft. of 3 inch schedule 40 pipe having internal diameter 7.79 cm, 8 threaded elbows, 2 globe valves, and 2 gate valves. The vapor pressure of water is 17.5 mm Hg.

    (a) What pump head and horsepower are required?
    (b) Would a pump whose generalized characteristic curve is $N_{pu} = 6.42 + 7.1\,N_Q + 913\,N_Q^2$ be suitable for the job? If so what impeller diameter, pump speed and motor horsepower should be used?
    (c) What is the maximum distance above the surface of the water in the well at which the pump can be located and still operate properly? The NPSH of the pump is given by NPSH = 0.3 + 246 Q. NPSH is in m water with Q in m^3/s.

    Viscosity of water = 1.002 cP.
    Density of water = 998 kg/m^3.

Bibliography

The material discussed in this chapter is discussed in many texts on fluid mechanics such as Sabersky *et al.* (1999).

Studies on the effect of surface roughness on the frictional resistance to flow through pipes were prompted originally by the observed decrease of flow in water mains over fairly long periods of time due to scaling of the pipelines. The generally accepted effects of pipe wall roughness are given by Colebrook and White (1937–38).

Details of the operating characteristics of centrifugal pumps are widely discussed in the literature and are available in specialized handbooks (Karassik *et al.* 2001). The treatment of the generalized characteristic curve given here is based on Stepanoff (1957), Labanoff and Ross (1992), Sulzer Brothers Ltd. (1989) and Grist (1998).

Wilson *et al.* (1997) discuss the performance of centrifugal pumps for slurry applications.

The derating of slurry pumps was investigated by Cave (1976).

References

Cave, I. (1976). Effects of Suspended Solids on the Performance of Centrifugal Pumps. *Hydrotransport 4* 4th international Conference on Hydraulic Transport of Solids in Pipes. BHRA Fluid Engineering. pp. H3–35–H3–52.

Colebrook, C.F. and White, C.M. (1937–8). The reduction of carrying capacity of pipes with age. *Journal of the Institution of Civil Engineers* **7**, 99–118.

Darby, R. (1996). *Chemical Engineering Fluid Mechanics*. Marcel Dekker.

Grist, E. (1998). *Cavitation and the Centrifugal Pump*. Taylor and Francis

Hooper, W.B. (1981). The two-K method to predicts pressure loss in fittings. *Chemical Engineering*, Aug. 24, 96–100.

Hooper, W.B. (1988). Calculate head loss caused by change in pipe size. *Chemical Engineering*, Nov. 7, 89–92.

Karassik, I.J., Messina, J.P., Cooper, P. and Heald, C.C. (2001). *Pump Handbook*, 3rd edition. McGraw-Hill.

Lobanoff, V.S. and Ross, R.R. (1992). *Centrifugal Pumps: Design and Applications*, 2nd edition. Gulf Publishing Company.

Sabersky, R.H., Acosta, A.J., Hauptmann, E.G. and Gates, E.M. (1999). *Fluid Flow: A First Course in Fluid Mechanics*. Prentice-Hall.

Stepanoff, A.J. (1957). *Centrifugal and Axial Flow Pumps*, 2nd edition. John Wiley and Sons.

Sulzer Brothers Ltd., (1989). *Sulzer Centrifugal Pump Handbook*. Elsevier Applied Science.

Wilson, K.C., Addie, G.R., Sellgren, A. and Clift, R. (1997). *Slurry Transport using Centrifugal Pumps*, 2nd edition. Blackie Academic and Professional.

3

Interaction between fluids and particles

3.1 Basic concepts

When a solid particle moves through a fluid it experiences a drag force that resists its motion. This drag force has its origin in two phenomena namely, the frictional drag on the surface and the increase in pressure that is generated in front of the particle as it moves through the fluid. The frictional drag is caused by the shearing action of the fluid as it flows over the surface of the particle. This component is called viscous drag.

A region of high pressure P_1 is formed immediately in front of the particle as it forces its way through the fluid. Likewise a region of relatively low pressure P_2 is formed immediately behind the particle in its wake. The pressure drop from the front of the particle to the rear is the result of the accumulated pressure exerted by the fluid integrated over the entire surface of the sphere. The pressure drop $P_1 - P_2$ gives rise to a force on the particle given by $(P_1 - P_2)A_c$ where A_c is the cross-sectional area of the particle measured perpendicular to the direction of motion. This is called the form drag. The total force on the particle is the sum of the viscous drag and the form drag (see Figure 3.1).

Both the form drag and viscous drag vary with the relative velocity between particle and fluid and with the density of the fluid. Many experiments have revealed that particles of different size show very similar behavior patterns when moving relative to the surrounding fluid be it air, water or any other viscous fluid. If the dimensionless groups

$$C_D = \frac{2 \times \text{Dragforce on particle}}{\text{Cross-sectional area} \times \rho_f \times \nu^2} = \frac{2F_D}{A_c\,\rho_f\,\nu^2} \tag{3.1}$$

and

$$Re_p = \frac{d_p\,\nu\,\rho_f}{\mu_f} \tag{3.2}$$

are evaluated at any relative velocity ν for any particle of size d_p, then all experimental data are described by a single relationship between C_D and Re_p. Results from a large number of experimental studies are summarized in Figure 3.2. The data points shown in Figures 3.2, 3.3 and 3.4 do not represent individually measured data points but are average values calculated by Lapple and Shepherd in 1940. They are included in the figures to provide points of reference to judge the adequacy of the fit of various empirical correlations to available data. C_D is called the drag coefficient of the particle and Re_p the particle Reynolds number. In spite

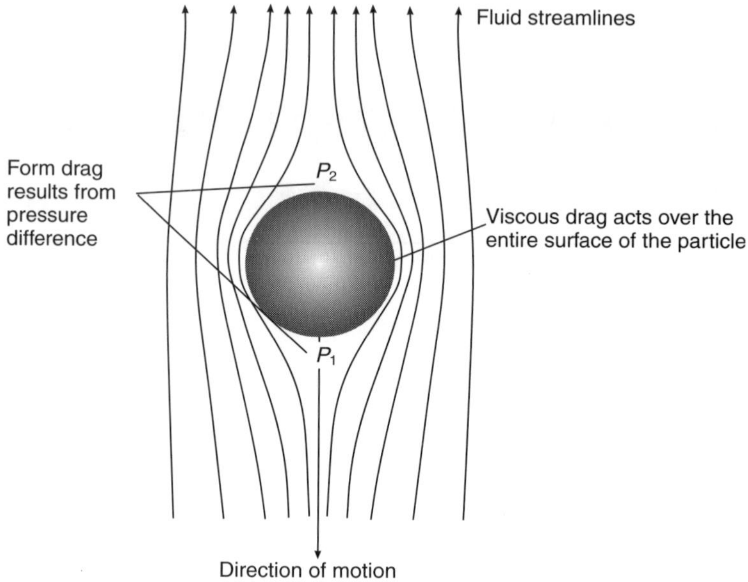

Figure 3.1 Streamlines that form around a particle that moves slowly through a fluid in the direction shown

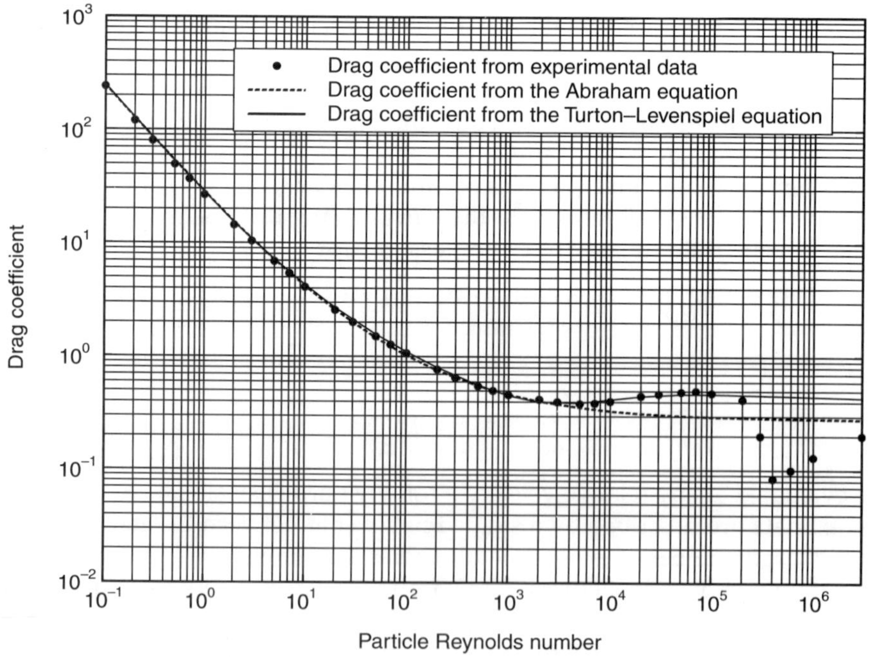

Figure 3.2 Drag coefficient of solid spheres plotted against the particle Reynolds number

of the apparent simplicity of the flow pattern that surrounds a moving sphere, it is not possible to derive a relationship between C_D and Re_p from fundamental fluid mechanical principles. The precise details of the flow field close to the particle are simply too complex. The only exception is the situation when the particle is spherical and the Reynolds number is very small when a completely analytical solution is available. This is called the Stokes regime and the main result is discussed in Section 3.2.2. However, the importance of the C_D–Re_p relationship for the analysis of a variety of practical problems has encouraged many authors to develop empirical relationships to fit the measured experimental data. In spite of their empirical nature, these relationships are of considerable utility in the solution of practical problems. A few of the many empirical drag coefficient correlations that have appeared in the literature are discussed in this chapter. These have been chosen because they prescribe adequate descriptions of the experimental data and provide convenient computational methods.

In the region $Re_p < 2 \times 10^3$ this data is described quite accurately by the Abraham equation

$$C_D = 0.28\left(1 + \frac{9.06}{Re_p^{1/2}}\right)^2 \tag{3.3}$$

In the region $Re_p < 2 \times 10^5$ the data is described well by the Turton–Levenspiel equation

$$C_D = \frac{24}{Re_p}\left(1 + 0.173\,Re_p^{0.657}\right) + \frac{0.413}{1 + 16300\,Re_p^{-1.09}} \tag{3.4}$$

The Abrahams and Turton–Levenspiel equations are plotted in Figure 3.2 and it is clear from this figure that the Abraham equation should not be used if $Re_p > 2 \times 10^3$ because it diverges considerably from the experimental data. The Turton–Levenspiel equation does a good job of representing the data up to $Re_p = 10^5$ above which the hydrodynamic field around the sphere becomes extremely complex. None of the applications that are discussed in this book will generate values of Re_p as high as 10^5 so this represents a practical upper limit for our purposes. Equations of the general form of equation 3.4 are frequently used and we shall refer to them as equations of Clift–Gauvin type in recognition of the authors who originally proposed their use.

Two alternative representations of the data are useful in practice and these are illustrated in Figures 3.3 and 3.4. Instead of using Re_p as the independent variable, the data is recomputed and the drag coefficient is plotted against the dimensionless groups

$$\Phi_1 = C_D\,Re_p^2 \tag{3.5}$$

and

$$\Phi_2 = \frac{Re_p}{C_D} \tag{3.6}$$

as independent variables as shown in Figures 3.3 and 3.4 respectively. When the data are plotted in this way they are seen to follow functional forms of

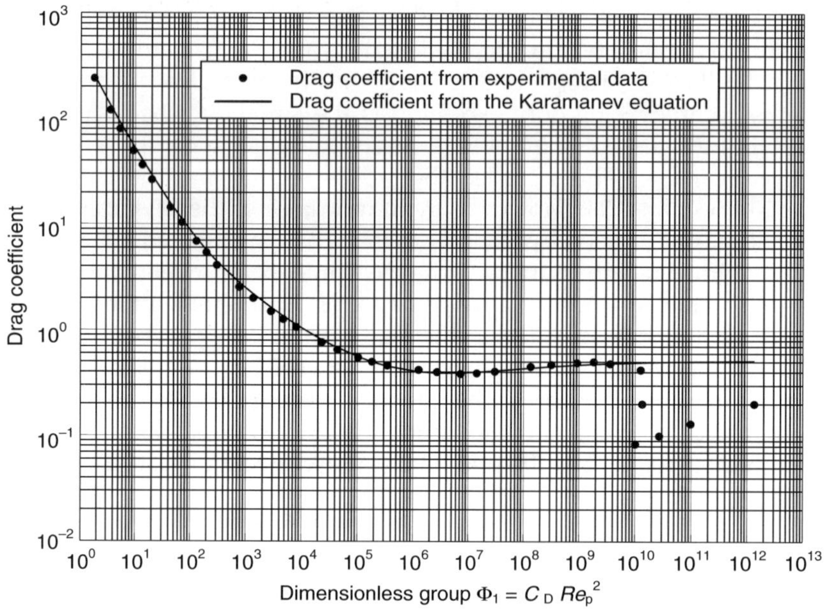

Figure 3.3 Drag coefficient for solid spheres plotted against the dimensionless group $\Phi_1 = C_D Re_p^2$. The line was plotted using equation 3.7. Use this graph to calculate the drag coefficient at terminal settling velocity when the particle size is known

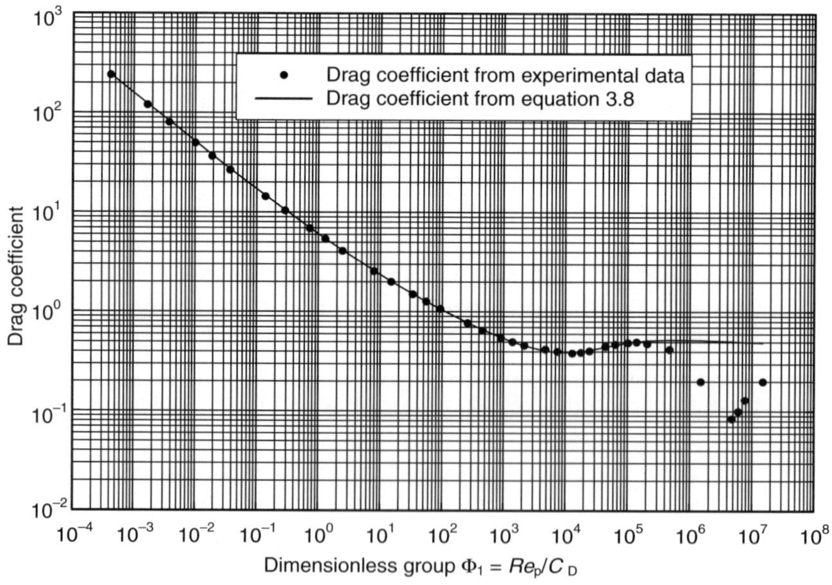

Figure 3.4 Drag coefficient plotted against the dimensionless group Re_p/C_D. The line is plotted using equation 3.8. Use this graph to calculate the drag coefficient at terminal settling velocity of a particle of unknown size when the terminal settling velocity is known

Clift–Gauvin type. This was pointed out by Karamanev (1996) who used the equation

$$C_D = \frac{432}{\Phi_1}\left(1 + 0.0470\,\Phi_1^{2/3}\right) + \frac{0.517}{1 + 154\,\Phi_1^{-1/3}} \tag{3.7}$$

to represent the data in Figure 3.3. An equation of similar form describes the data in Figure 3.4

$$C_D = \frac{4.90}{\Phi_2^{1/2}}\left(1 + 0.243\,\Phi_2^{1/3}\right) + \frac{0.416}{1 + 3.91 \times 10^4\,\Phi_2^{-1}} \tag{3.8}$$

Depending on the problem context, the drag coefficient can be calculated from equations 3.3, 3.4, 3.7 or 3.8 and then used with equation 3.1 to calculate the force on the particle as it moves through the fluid.

$$\text{Force} = \frac{C_D}{2}\rho_f\,\nu^2\,A_c \tag{3.9}$$

3.2 Terminal settling velocity

If a particle falls under gravity through a viscous fluid it will accelerate for a short while but as the particle moves faster the drag force exerted by the fluid increases until the drag force is just equal to the net gravitational force less the buoyancy that arises from the immersion of the particle in the fluid. When these forces are in balance the particle does not accelerate any further and it continues to fall at a constant velocity. This condition is known as terminal settling.

The terminal settling velocity ν_T can be evaluated by balancing the drag and buoyancy forces

$$\nu_p(\rho_s - \rho_f)g = \frac{C_D^*}{2}\rho_f\,\nu_T^2\,A_c \tag{3.10}$$

where ν_p is the volume of the particle. C_D^* is the drag coefficient at terminal settling velocity.

3.2.1 *Settling velocity of an isolated spherical particle*

When the particle is spherical, the geometrical terms in equation 3.10 can be written in terms of the particle diameter

$$\frac{\pi}{6}d_p^3(\rho_s - \rho_f)g = \frac{C_D^*}{2}\rho_f\,\nu_T^2\,\frac{\pi}{4}d_p^2 \tag{3.11}$$

and the drag coefficient of a spherical particle at terminal settling velocity is given by

$$C_D^* = \frac{4}{3}\frac{(\rho_s - \rho_f)}{\rho_f \, \nu_T^2} g \, d_p \tag{3.12}$$

The particle Reynolds number at terminal settling velocity is given by

$$Re_p^* = \frac{d_p \, \nu_T \, \rho_f}{\mu_f} \tag{3.13}$$

It is not possible to solve equation 3.11 directly because C_D is a function of both ν_T and the particle size d_p through the relationship shown in Figure 3.2 or that given by either of the Abraham or Turton–Levenspiel equations. Two different solution procedures are commonly required in practice: the calculation of the terminal settling velocity for a particle of given size or the calculation of the size of the spherical particle that has a prespecified terminal settling velocity. These problems can be solved without recourse to trial and error methods by considering the two dimensionless groups Φ_1 and Φ_2 both evaluated at the terminal settling velocity.

$$\begin{aligned} \Phi_1^* = C_D^* \, Re_p^{*2} &= \frac{4}{3}\frac{\rho_s - \rho_f}{\rho_f}\frac{g d_p}{\nu_T^2}\left(\frac{d_p \, \nu_T \, \rho_f}{\mu_f}\right)^2 \\ &= \left[\frac{4}{3}(\rho_s - \rho_f)\,\rho_f \frac{g}{\mu_f^2}\right] d_p^3 = d_p^{*3} \end{aligned} \tag{3.14}$$

and

$$\Phi_2^* = \frac{Re_p^*}{C_D^*} = \left[\frac{3}{4}\frac{\rho_f^2}{(\rho_s - \rho_f)\,\mu_f \, g}\right] \nu_T^3 = V_T^{*3} \tag{3.15}$$

d_p^* and V_T^* are called the dimensionless particle diameter and the dimensionless terminal settling velocity of a sphere respectively. d_p^* can be evaluated from a knowledge of the properties of the fluid and the size of the particle and d_p^* is independent of the terminal settling velocity. V_T^* is independent of the particle size and can be evaluated if the terminal settling velocity is known. Φ_1^* is often called the Archimedes number which is usually represented by the symbol *Ar*.

From the definition of Φ_1^* and Φ_2^* given in equations 3.14 and 3.15, the following relationships are easily derived.

$$Re_p^* = (\Phi_1^* \, \Phi_2^*)^{1/3} = d_p^* \, V_T^* \tag{3.16}$$

$$C_D^* = \frac{Re_p^*}{\Phi_2^*} = \frac{\Phi_1^*}{Re_p^{*2}} = \frac{d_p^*}{V_T^{*2}} \tag{3.17}$$

$$Re_p^* = C_D^* \, V_T^{*3} = \left(\frac{d_p^{*3}}{C_D^*}\right)^{1/2} \tag{3.18}$$

These are all equivalent definitions of the terminal settling condition.

The drag coefficient at terminal settling velocity can be calculated easily from Figure 3.3 or equation 3.7 if the dimensionless diameter of the particle is known. It is then a simple matter to evaluate the terminal settling velocity from equation 3.11 or more conveniently from equation 3.17.

The diameter of the particle that has known terminal settling velocity can be calculated using a similar procedure. In this case, the dimensionless settling velocity V_T^* is calculated using equation 3.15 and the drag coefficient at terminal settling velocity can then be calculated from equation 3.8 or read from Figure 3.4. In either case, the required variable can be calculated directly from equation 3.11 once C_D^* is known or more conveniently from equation 3.17.

If the Abraham equation is used to describe the relationship between C_D and Re_p the method described above leads to a completely analytical solution. Using equations 3.16 and 3.15 and substituting for C_D^* and Re_p^* gives

$$V_T^{*3} = \frac{Re_p^*}{C_D^*} = \frac{d_p^* V_T^*}{0.28\left(1 + \dfrac{9.06}{(d_p^* V_T^*)^{1/2}}\right)^2} \tag{3.19}$$

$$1 + \frac{9.06}{(d_p^* V_T^*)^{1/2}} = \frac{d_p^{*1/2}}{0.28^{1/2} V_T^*} \tag{3.20}$$

This equation can be re-arranged to show more clearly the relationship between V_T^* and d_p^*

$$0.28^{1/2} V_T^* d_p^{*1/2} + 9.06 \times 0.28^{1/2} V_T^{*1/2} - d_p^* = 0 \tag{3.21}$$

This equation can be solved for V_T^* in terms of d_p^* or for d_p^* in terms of V_T^*. Either way the solution procedure is straight forward because equation 3.21 is quadratic in the variables $V_T^{*1/2}$ and $d_p^{*1/2}$.

The solutions are

$$V_T^* = \frac{20.52}{d_p^*}\left[\left(1 + 0.0921\, d_p^{*3/2}\right)^{1/2} - 1\right]^2 \tag{3.22}$$

or

$$d_p^* = 0.070\left(\left(1 + \frac{68.49}{V_T^{*3/2}}\right)^{1/2} + 1\right)^2 V_T^{*2} \tag{3.23}$$

The drag coefficient at terminal settling velocity is related to d_p^* and V_T^* most directly by equation 3.17. These solutions are due to Concha and Almendra (1979). This approach to the analysis of terminal settling velocity provides straight forward linear procedures for the calculation of terminal settling velocity or particle size. No inconvenient iterative calculations are required. The methods are summarized in Table 3.1. These calculation methods are

Table 3.1 Summaries of linear calculation procedures for terminal settling velocity or particle size.

Basis	*Method*
Case 1: Particle size is known, terminal settling velocity is required	
Karamanev equation	$d_p \rightarrow$ Eqn 3.14 $\rightarrow \Phi_1^* \rightarrow$ Eqn 3.7 $\rightarrow C_D^* \rightarrow$ Eqn 3.17 $\rightarrow Re_p^* \rightarrow$ Eqn 3.2 $\rightarrow v_T$
Abraham equation	$d_p \rightarrow$ Eqn 3.14 $\rightarrow d_p^* \rightarrow$ Eqn 3.22 $\rightarrow V_T^* \rightarrow$ Eqn 3.16 $\rightarrow Re_p^* \rightarrow$ Eqn 3.2 $\rightarrow v_T$
Case 2: Terminal settling velocity is known, particle size is required	
Equation 3.8	$v_T \rightarrow$ Eqn 3.15 $\rightarrow \Phi_2^* \rightarrow$ Eqn 3.8 $\rightarrow C_D^* \rightarrow$ Eqn 3.17 $\rightarrow Re_p^* \rightarrow$ Eqn 3.2 $\rightarrow d_p$
Abraham equation	$v_T \rightarrow$ Eqn 3.15 $\rightarrow V_T^* \rightarrow$ Eqn 3.23 $\rightarrow d_p^* \rightarrow$ Eqn 3.16 $Re_p^* \rightarrow$ Eqn 3.2 $\rightarrow d_p$

sufficiently straight forward to be undertaken using a simple calculator. However they are tiresome and they are available in the FLUIDS computational toolbox for convenience. Illustrative examples 3.1 and 3.2 demonstrate their application.

Illustrative example 3.1 Terminal settling velocity

Calculate the terminal settling velocity of a glass sphere of diameter 0.1 mm in a fluid having density 982 kg/m^3 and viscosity 0.0013 kg/ms. Density of glass is 2820 kg/m^3. Also calculate the drag coefficient at terminal settling velocity.

Solution

Since the particle size is known, calculate d_p^*.

$$\Phi_1^* = d_p^{*3} = \frac{4}{3}\frac{(\rho_s - \rho_f)\,\rho_f\, g\, d_p^3}{\mu_f^2}$$

$$= \frac{4}{3}\frac{(2820 - 982)982 \times 9.81 \times (0.1 \times 10^{-3})^3}{(0.0013)^2}$$

$$= 13.97$$

Using the Karamanev equation,

$$C_D^* = \frac{432}{\Phi_1^*}(1 + 0.0470\,\Phi_1^{*2/3}) + \frac{0.517}{1 + 154\,\Phi_1^{-1/3}}$$

$$= 39.355 + 0.008 = 39.36$$

Check this on the graph of C_D vs Φ_1 (Figure 3.3)

$$Re_p^* = \left(\frac{\Phi_1^*}{C_D^*}\right)^{1/2} = \left(\frac{13.97}{39.36}\right)^{1/2} = 0.596$$

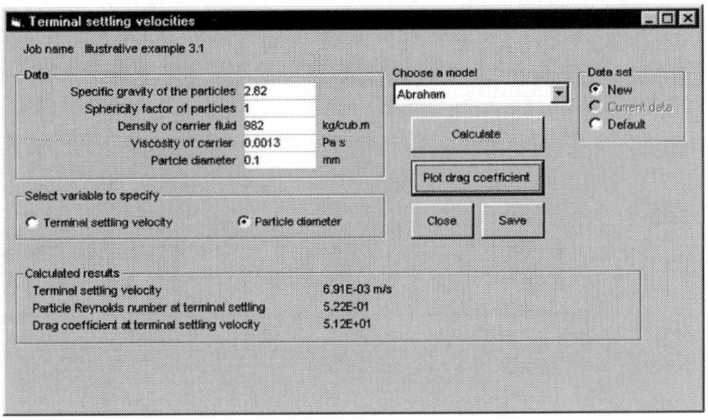

Figure 3.5 FLUIDS toolbox data screen to calculate the terminal settling velocity of a particle of known diameter. The Abraham model is chosen

Check this on the C_D vs Re_p graph (Figure 3.2)

$$\nu_T = Re^*_p \frac{\mu_f}{d_p\,\rho_f} = \frac{0.596 \times 0.0013}{0.1 \times 10^{-35} \times 982} = 7.89 \times 10^{-3}\,\text{m/s}$$

Alternatively use the Concha–Almendra procedure

$$d^*_p = (13.97)^{1/3} = 2.408$$

$$V^*_T = \frac{20.52}{d^*_p}[(1 + 0.0921\; d_p^{*3/2})^{1/2} - 1]^2$$

$$= 0.217$$

$$Re^*_p - d^*_p V^*_T = 2.408 \times 0.217 = 0.522$$

$$V_T = \frac{Re^*_p \mu_f}{d_p \rho_f} = \frac{0.522 \times 0.0013}{0.1 \times 10^{-3} \times 982} = 6.91 \times 10^{-3}\,\text{m/s}$$

The drag coefficient at terminal settling velocity is calculated from

$$C^*_D = \frac{d^*_p}{V^{*2}_T} = \frac{2.408}{0.217^2} = 51.1$$

These calculations can be checked easily using the terminal settling velocity calculated in the FLUIDS toolbox as shown in Figure 3.5. Note the different values obtained for the terminal settling velocity using these two methods. These differences result from the differences between the Abraham and Karamanev equations for the drag coefficient.

Illustrative example 3.2 Size of settling particle

Find the size of a spherical particle of glass that has a terminal settling velocity of 7.4 mm/s in a fluid of density 982 kg/m^3 and viscosity 0.0013 kg/ms. Density of glass is 2820 kg/m^3.

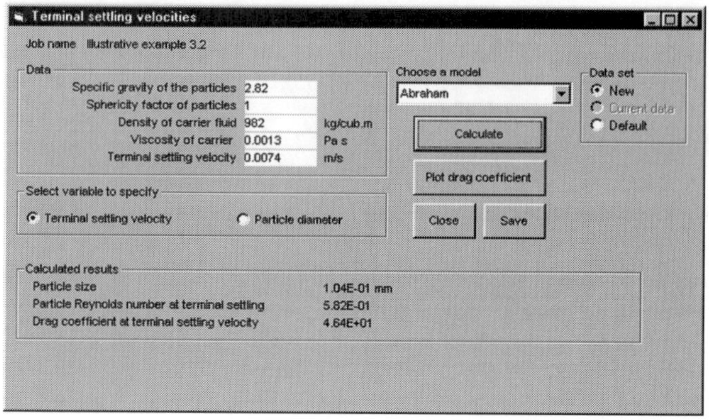

Figure 3.6 FLUIDS toolbox window to calculate the size of a particle with known terminal settling velocity. The Abraham equation is chosen

Solution

Since the terminal settling velocity is known, start the calculation with V_T^*.

$$\Phi_2^* = V_\mathrm{T}^{*3} = \frac{3}{4}\frac{\rho_\mathrm{f}^2}{(\rho_\mathrm{s} - \rho_\mathrm{f})\,\mu_\mathrm{f}\, g}\nu_\mathrm{T}^3$$

$$= \frac{3}{4}\frac{982^2 \times (7.4 \times 10^{-3})^3}{(2820 - 982) \times 0.0013 \times 9.81} = 0.0125$$

$$C_\mathrm{D}^* = \frac{4.90}{\Phi_2^{*1/2}}(1 + 0.243\,\Phi_2^{*1/3}) + \frac{0.416}{1 + 3.91 \times 10^4\,\Phi_2{}^{*-1}}$$

$$= 46.29 + 1.33 \times 10^{-7} = 46.29$$

$$Re_\mathrm{p}^* = C_\mathrm{D}^*\, V_\mathrm{T}^{*3} = 46.29 \times 0.0125 = 0.579$$

$$d_\mathrm{p} = \frac{Re_\mathrm{p}^* \times \mu_\mathrm{f}}{\nu_\mathrm{T} \times \rho_\mathrm{f}} = \frac{0.579 \times 0.0013}{7.4 \times 10^{-3} \times 982} = 1.04 \times 10^{-4}\,\mathrm{m}$$

Alternatively, the Conca–Almendra method can be used.

$$d_\mathrm{p}^* = 0.070\left[\left(1 + \frac{68.49}{V_\mathrm{T}^{*3/2}}\right)^{1/2} + 1\right]^2 V_\mathrm{T}^{*2} = 2.504$$

$$Re_\mathrm{p}^* = d_\mathrm{p}^*\, V_\mathrm{T}^* = 2.504 \times (0.0125)^{1/3} = 0.581$$

$$d_\mathrm{p} = \frac{Re_\mathrm{p}^* \times \mu_\mathrm{f}}{\nu_\mathrm{T} \times \rho_\mathrm{f}} = 1.04 \times 10^{-4}\,\mathrm{m}$$

These calculations can be checked using the FLUIDS software toolbox as shown in Figure 3.6.

The reader should note that equations 3.7 and 3.8 are not exactly equivalent to the Turton–Levenspiel equation 3.4. Consequently, these equations give slightly different values for the drag coefficient. Equations 3.7 and 3.8 should be used only when approximate manual calculations are undertaken. They are not used in the FLUIDS toolbox.

3.2.2 Stokes' Law

When the relative velocity between a sphere and the fluid is very low, the drag force is due entirely to surface drag and the relationships that define the interaction between the fluid and the particle can be calculated exactly from the basic differential equations that describe the velocity field around the sphere. The solution can be summarized by the equation

$$F_D = 3\pi\mu\, d_p\, v \tag{3.24}$$

Substitution in equation 3.1

$$C_D = \frac{24}{Re_p} \tag{3.25}$$

The terminal settling is obtained by substitution of this equation into equation 3.10

$$\frac{\pi}{6} d_p^3 (\rho_s - \rho_f) g = \frac{24}{2\, Re_p} \rho_f\, v_T^2 \frac{\pi}{4} d_p^2 \tag{3.26}$$

$$v_T = \frac{\rho_s - \rho_f}{18\, \mu_f} g d_p^2 \tag{3.27}$$

This is known as Stokes' law of settling. Stokes law is particularly useful for the analysis of the motion of very small particles. It should not be used if the particle Reynolds number is greater than 0.01.

Illustrative example 3.3

Use Stokes' Law to calculate the terminal settling velocity of a glass sphere of diameter 0.1 mm in a fluid having density 982 kg/m^3 and viscosity 0.0013 Pa s. Density of glass is 2820 kg/m^3.

Solution

Stokes' Law

$$\begin{aligned} V_T &= \frac{\rho_s - \rho_f}{18\, \mu_f} g\, d_p^2 \\ &= \frac{2820 - 982}{18 \times 0.0013} 9.81 (0.1 \times 10^{-3})^2 \\ &= 7.71 \times 10^{-3}\ \text{m/s} \end{aligned}$$

Calculate the Reynolds number at terminal settling velocity to establish whether Stokes' Law can be applied.

$$Re_{\mathrm{p}}^{*} = \frac{d_{\mathrm{p}}\, V_{\mathrm{T}}\, \rho_{\mathrm{f}}}{\mu_{\mathrm{f}}}$$

$$= \frac{0.1 \times 10^{-3} \times 7.71 \times 10^{-3} \times 9.82}{0.0013} = 0.582$$

This does not satisfy the condition $Re_{\mathrm{p}} < 0.01$ for the application of Stokes' Law.

3.3 Isolated isometric particles of arbitrary shape

In practical engineering situations it is more common to work with particles of irregular shape than it is to work with spherical particles. The shape of a particle is determined by its method of manufacture and these vary widely in the process industries. Common synthesis processes include precipitation, crystallization, deposition, comminution and agglomeration. These processes invariably produce non-spherical particles unless they are designed expressly to produce spheres. Deviations from spherical shape have a significant effect on the drag coefficient and therefore on the terminal settling velocity of the particles. The methods described in Section 3.2 must be modified in order to calculate drag coefficients and therefore terminal settling velocities of non-spherical particles.

Irregular shape influences the motion of a sphere in a fluid in a number of ways. The surface drag characteristics are changed and usually increase so that the irregular particles have lower terminal settling velocities than spheres. Irregularities in shape can lead to oscillatory and erratic relative motion between the particle and the fluid. When the particles are not isometric and have significantly different dimensions in three orthogonal directions, the orientation of the particle relative to the direction of motion has a profound effect on the drag coefficient. If the particle has internal surfaces resulting from holes such as washers or porous agglomerates, the drag coefficient will be affected and in most cases accurate estimates of C_{D} are difficult to obtain.

When the particles are not spherical but at least isometric (i.e. they have equal or approximately equal dimensions in three orthogonal directions) useful and accurate correlations have been developed. There are two factors that make the non-spherical particles differ from their spherical counterparts: the surface area per unit volume is larger giving rise to larger surface drag and the irregular shape of the particle gives rise to oscillatory and vibratory motions as the particle settles through the fluid.

Two methods have been used to calculate the drag coefficients and terminal settling velocities of particles that are approximately isometric but not spherical. The first relies on empirical adjustments to the drag coefficient equations to account for particle shape and the second relies on empirical modifications to the particle Reynolds number and the drag coefficient so that the correlations for spherical particles can be used. Both methods are described in the following sections.

In both cases, a quantitative measure of the particle shape and size must be defined and used consistently. This is not straightforward because there is no unique definition of the size or shape of an non-spherical particle. Fortunately, most authors have standardized on two definitions that have found almost universal acceptance so that different methods can be compared. The particle size used is the equivalent volume diameter.

$$d_e = \left(\frac{6\,v_p}{\pi}\right)^{1/3} \tag{3.28}$$

which is the diameter of the sphere having the same volume as the particle. d_e is used to calculate the particle Reynolds number for particles of any shape. The shape is quantitatively specified by the sphericity of the particle which is defined as

$$\begin{aligned}\psi &= \frac{\text{Surface area of sphere with equal volume}}{\text{Surface area of particle}} \\ &= \frac{\pi\,d_e^2}{a_p}\end{aligned} \tag{3.29}$$

Both d_e and ψ can be calculated when the particle has a regular shape but is not spherical such as a cube or regular octahedron. When the particles have irregular shapes these calculations are impossible and both d_e and ψ must be measured experimentally. These measurements require special techniques and image analysis using optical or electron microscopy is a reliable and accurate method that has been developed during the past 20 years. The image analysis method does not generate the surface area or volume of a single isolated particle but average values for many particles in a population. This restriction is not limiting for most industrial processes which usually operate with large numbers of particles.

The image analysis method is implemented in the following manner. A representative sample of the particle population is mounted in epoxy resin in such a way that the particles are oriented randomly and uniformly. This mount is then sectioned and the section surface is polished flat. This leaves a random selection of 2D sections through the particles exposed on the surface. Digital images of these polished sections are then generated in the optical or scanning electron microscope. Using appropriate image analysis software the necessary geometric properties of the each particle in the section can be accurately measured. In particular, the total perimeter of the particle sections divided by the total area of the sections gives an unbiased estimate of the perimeter per unit area of the solid particles.

$$B_A = \frac{B}{A} \tag{3.30}$$

where B is the total measured perimeter and A is the total particle section area. A theorem of stereology gives the average surface area per unit volume of the particles in the population as

$$S_V = \frac{4}{\pi} B_A \tag{3.31}$$

Equation 3.31 is exact and does not depend on the shape of the particles. The sphericity is related to S_V by

$$\psi = \frac{6}{S_V d_e} \tag{3.32}$$

If the particles are large enough to be handled individually their volumes can be measured directly by fluid displacement pycnometry. Water displacement can be used satisfactorily but helium displacement pycnometry is now commonly used and provides good accuracy. When the particles are smaller than about 1 mm, pycnometry is not a viable technique and microscopic stereology is used.

3.3.1 Empirical modifications to drag coefficient correlations

One approach to dealing with non-spherical particles is to find empirical relationships for the coefficients in the drag coefficient vs Re_p correlation (see Figure 3.7). A Clift–Gauvin type equation can be used

$$C_D = \frac{24}{Re_p}\left(1 + ARe_p^B\right) + \frac{C}{1 + DRe_p^{-1}} \tag{3.33}$$

and the parameters *A*, *B*, *C* and *D* are related to the sphericity as shown in Table 3.2.

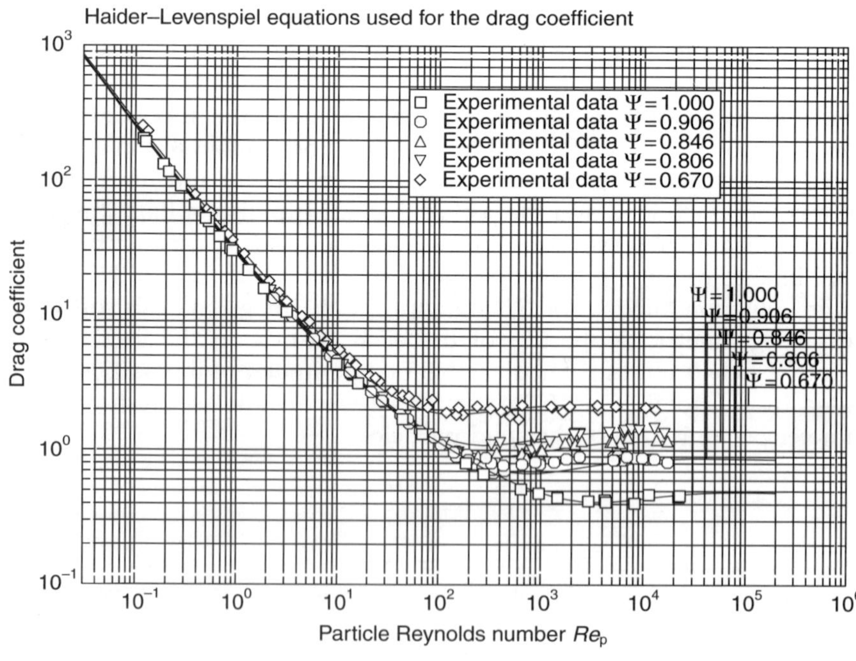

Figure 3.7 Drag coefficient for isometric non-spherical particles using the Haider–Levenspiel correlation compared to data of Pettyjohn and Christiansen (1948)

Table 3.2 Drag coefficient parameters for non-spherical particles of regular shape

Shape	ψ	*A*	*B*	*C*	*D*
Sphere	1.000	0.1806	0.6459	0.4251	6880.95
Cubic octahedron	0.906	0.2155	0.6028	0.8203	1080.84
Octahedron	0.846	0.2559	0.5876	1.2191	1154.13
Cubes	0.806	0.2734	0.5510	1.406	762.39
Tetrahedron	0.670	0.4531	0.4484	1.945	101.18

These values are interpolated using the following polynomials to obtain values at other values of the sphericity ψ.

$$\begin{aligned} A &= \exp(2.3288 - 6.4581\psi + 2.4486\,\psi^2) \\ B &= 0.0964 + 0.5565\psi \\ C &= \exp(4.905 - 13.8944\psi + 18.422\,\psi^2 - 10.2599\,\psi^3) \\ D &= \exp(1.4681 + 12.2584\psi - 20.7322\,\psi^2 + 15.8855\,\psi^3) \end{aligned} \tag{3.34}$$

These correlations are due to Haider and Levenspiel (1989).

The drag coefficient calculated using equation 3.33 with the constants given by equation 3.34 is plotted against Re_p in Figure 3.7 where it is compared to the experimental data of Pettyjohn and Christiansen (1948). C_D is plotted against Φ_1

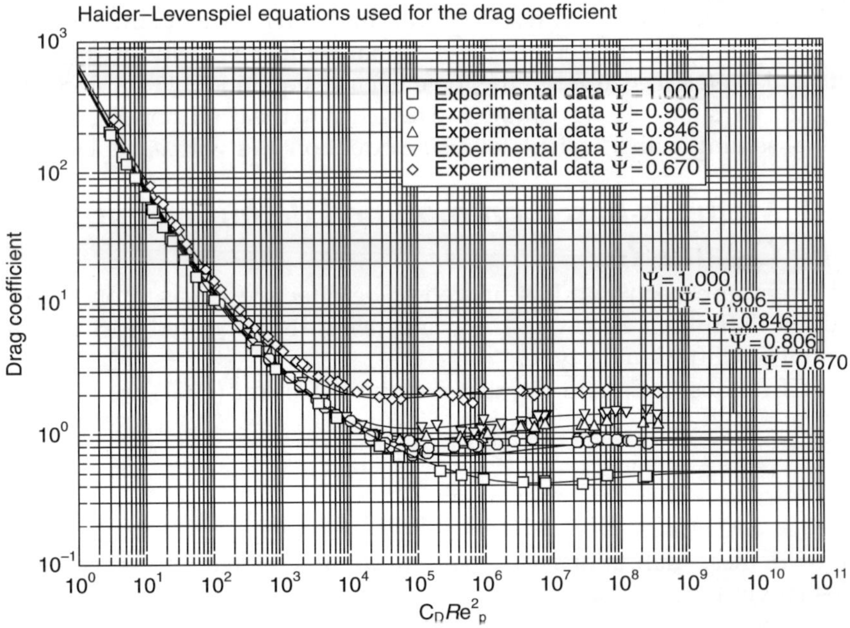

Figure 3.8 Drag coefficient for isometric non-spherical particles using the Haider–Levenspiel correlation compared to data of Pettyjohn and Christiansen (1948)

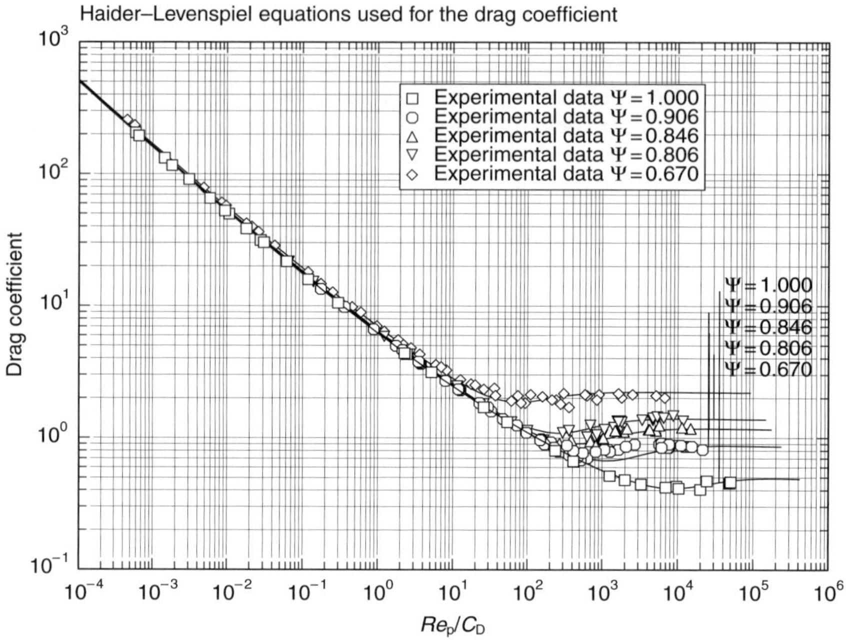

Figure 3.9 Drag coefficient for isometric non-spherical particles using the Haider–Levenspiel correlation compared to data of Pettyjohn and Christiansen (1948)

and Φ_2 in Figures 3.8 and 3.9 respectively which can be used in the same way as Figures 3.3 and 3.4 to calculate the terminal settling velocity or the effective particle diameter. d_e is used in the force balance equation 3.11 in place of d_p to define the terminal settling condition. This leads to the condition $Re_p^* = d_e^* V_T^*$ for terminal settling with d_e^* calculated using equation 3.14 with d_e replacing d_p.

3.3.2 Methods based on modified definitions of C_D and Re_p

A different approach that has proved to be effective is based on the use of modified definitions of C_D and Re_p and using the modified variables in an equation of Clift–Gauvin type.

Modifications to C_D and Re_p that are due to Ganser (1993) are

$$Re_M = Re_p K_1 K_2 \tag{3.35}$$

$$C_{D\,M} = \frac{C_D}{K_2} \tag{3.36}$$

and with these definitions C_{DM} and Re_{pM} are related by an equation of Clift–Gauvin type

$$C_{DM} = \frac{24}{Re_M}\left(1 + 0.1118\, Re_M^{0.6567}\right) + \frac{0.4305}{1 + 3305(Re_M)^{-1}} \tag{3.37}$$

The parameters K_1 and K_2 are related to the sphericity as follows

$$K_1 = \left(\frac{1}{3} + \frac{2}{3\,\psi^{0.5}}\right)^{-1} \tag{3.38}$$

and

$$K_2 = 10^{1.8148(-\log_{10}\psi)^{0.5743}} \tag{3.39}$$

These correlations can be used to generate graphs of the drag coefficient equivalent to Figures 3.7, 3.8 and 3.9. The reader is referred to the terminal velocity section of the FLUIDS computational toolbox to generate these graphs.

Other modifications to the drag coefficient and the particle Reynolds number are used and two due to Concha and Barrientos (1986) are

$$C_{D\,M} = \frac{C_D}{f_A(\psi) f_C(\lambda)} \tag{3.40}$$

and

$$Re_M = \frac{Re_p}{[f_B(\psi) f_D(\lambda)]^2} \tag{3.41}$$

In these equations λ is the density ratio

$$\lambda = \frac{\rho_s}{\rho_f} \tag{3.42}$$

The functions f_A, f_B, f_c and f_D account for the effect of sphericity and density ratio on the drag coefficient and the particle Reynolds number. These functions have been chosen so that the modified drag coefficient is related to the modified Reynolds number using the same equation that describes the drag coefficient for spherical particles.

The empirical functions are given by

$$f_A(\psi) = \frac{5.42 - 4.75\psi}{0.67} \tag{3.43}$$

$$f_B(\psi) = \left(0.843 f_A(\psi) log \frac{\psi}{0.065}\right)^{-1/2} \tag{3.44}$$

$$f_C(\lambda) = \lambda^{-0.0145} \tag{3.45}$$

$$f_D(\lambda) = \lambda^{0.00725} \tag{3.46}$$

The modified variables satisfy the spherical drag coefficient equation. Thus the Abraham equation for non-spherical particles is

$$C_{D\,M} = 0.28\left(1 + \frac{9.06}{Re_M^{1/2}}\right)^2 \tag{3.47}$$

or equations of Clift–Gauvin type become

$$C_{DM} = \frac{24}{Re_M}\left(1 + A\, Re_M^B\right) + \frac{C}{1 + D\, Re_M^{-1}} \tag{3.48}$$

Note that

$$f_A(1) = f_B(1) = f_C(1) = f_D(1) = 1.0 \tag{3.49}$$

so the modified equation correctly describes the behavior of spherical particles.

It is possible to extend this idea of parameter normalization so that a single relation between the dimesionless particle size and the dimensionless terminal settling velocity can describe the drag behavior of particles of any shape. Extensions due to Concha and Barrientos (1986) can be used to define modified dimensionless particle size and dimensionless settling velocity as follows

$$d_{eM}^* = d_e^* \left(\beta(\psi)\right)^{2/3} \left(\eta(\lambda)\right)^{2/3} \tag{3.50}$$

and

$$V_M^* = \frac{V_T^*}{\alpha(\psi)\gamma(\lambda)\beta(\psi)^{2/3}\,\eta(\lambda)^{2/3}} \tag{3.51}$$

where d_e^* and V_T^* are evaluated from equations 3.14 and 3.15 using d_e rather than d_p in equation 3.14. The extended functions α, β, γ and η are related to f_A, f_B, f_c and f_D as follows;

$$\alpha(\psi) = f_B^2(\psi) \tag{3.52}$$

$$\beta(\psi) = \left(f_A^{1/2}(\psi) f_B^2(\psi)\right)^{-1} \tag{3.53}$$

$$\gamma(\lambda) = f_D^2(\lambda) \tag{3.54}$$

$$\eta(\lambda) = \left(f_c(\lambda)^{1/2} f_D(\lambda)^2\right)^{-1} \tag{3.55}$$

With these definitions of $\alpha(\psi)$, $\beta(\psi)$, $\gamma(\psi)$ and $\eta(\psi)$, it is easy to show that the modified variables satisfy the relationships 3.16 and 3.17 at terminal settling velocity.

$$d_{eM}^*\, V_M^* = d_e^*\, \beta^{2/3}\, \eta^{2/3} \frac{V_T^*}{\alpha\gamma\, \beta^{2/3}\, \eta^{2/3}} = \frac{d_e^*\, V_T^*}{\alpha\gamma} = \frac{Re_P^*}{f_B^2 f_D^2} = Re_M^* \tag{3.56}$$

and

$$\frac{Re_M^*}{C_{DM}^*} = \frac{Re_P^*}{f_B^2 f_D^2}\frac{f_A f_C}{C_D^*} = V_T^* \frac{f_A f_C}{f_B^2 f_D^2} = V_M^{*3} \tag{3.57}$$

This leads to an explicit solution of the modified Abraham equation in the same way as for spherical particles to give

$$V_M^* = \frac{20.52}{d_{eM}^*}\left[\left(1 + 0.0921\, d_{eM}^{*3/2}\right)^{1/2} - 1\right]^2 \tag{3.58}$$

and

$$d_{\text{eM}}^{*} = 0.070\left(\left(1+\frac{68.49}{V_{\text{M}}^{*3/2}}\right)^{1/2}+1\right)^{2} V_{\text{M}}^{*2} \tag{3.59}$$

which are identical in form to equations 3.22 and 3.23.

Equations of the Clift–Gauvin type do not lead to a neat closed form solution but a convenient computational method can be developed using the drag coefficient plots based on the dimensionless groups $\Phi_{1\text{M}}$ and $\Phi_{2\text{M}}$, the modified counterparts of Φ_1 and Φ_2.

$$\Phi_{1\text{M}} = C_{\text{D M}}\, Re_{\text{M}}^{2} \tag{3.60}$$

and

$$\Phi_{2\text{M}} = \frac{Re_{\text{M}}}{C_{\text{D M}}} \tag{3.61}$$

$\Phi_{1\text{M}}$ and $\Phi_{2\text{M}}$ can be used with Figures 3.3 and 3.4 to obtain values of the drag coefficient at terminal settling velocity.

The application of these methods is illustrated in the following example.

Illustrative example 3.4

Calculate the terminal settling velocity of a glass cube having edge dimension 0.1 mm in a fluid of density 982 kg/m^3 and viscosity 0.0013 kg/ms. The density of the glass is 2820 kg^3. Calculate the equivalent volume diameter and the sphericity factor.

$$d_{\text{e}} = \left(\frac{6v_{\text{b}}}{\pi}\right)^{1/3} = \left(\frac{6\times 10^{-12}}{\pi}\right)^{1/3} = 1.241\times 10^{-4}\,\text{m}$$

$$\psi = \frac{\pi d_{\text{e}}^{2}}{a_{\text{p}}} = \frac{\pi(1.241\times 10^{-4})^{2}}{6\times 10^{-8}} = 0.806$$

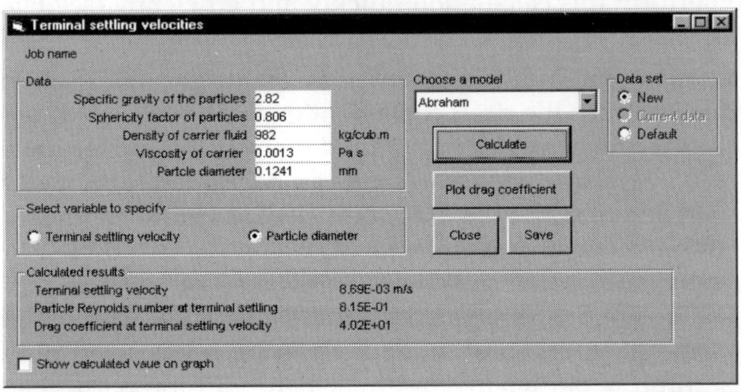

Figure 3.10 FLUIDS toolbox screen for calculation of terminal settling velocity in illustrative example 3.4

$$\lambda = \frac{\rho_s}{\rho_f} = \frac{2820}{982} = 2.872$$

$$f_A(\psi) = \frac{5.42 - 4.75\psi}{0.67} = 2.375$$

$$f_B(\psi) = \left(0.843 f_A(\psi) \log \frac{\psi}{0.065}\right)^{-1/2}$$

$$= 0.676$$

$$f_C(\lambda) = 0.985$$

$$\alpha(\psi) = f_B^2(\psi) = 0.457$$

$$\beta(\psi) = \left(f_A^{1/2}(\psi) f_B^2(\psi)\right)^{-1} = 1.421$$

$$\gamma(\psi) = f_D^2 = 1.015$$

$$\eta(\psi) = \left(f_C(\lambda)^{1/2} f_D(\gamma)^2\right)^{-1} = 0.992$$

$$d_e^* = \left[\frac{4}{3}(\rho_s - \rho_f)\rho_f \frac{g}{\mu_f^2}\right]^{1/3} d_e = 2.989$$

$$d_{eM}^* = d_e^*(\beta(\psi))^{2/3}(\eta(\lambda))^{2/3} = 3.757$$

$$V_M^* = \frac{20.52}{d_{eM}^*}\left[(1 + 0.0921 d_{eM}^{*3/2})^{1/2} - 1\right]^2$$

$$= 0.467$$

$$V_T^* = V_M^* \alpha(\psi)\gamma(\lambda)\beta(\psi)^{2/3}\eta(\lambda)^{2/3}$$

$$= 0.273$$

$$v_T = V_T^* \left[\frac{3}{4} \frac{\rho_f^2}{(\rho_s - \rho_f)\,\mu_f\, g}\right]^{-1/3} = 8.70 \times 10^{-3} \text{ m/s}$$

These calculations are straightforward but tedious. The software toolbox can be used to perform this calculation quickly and efficiently (see Figure 3.10).

An alternative graphical representation of the terminal settling velocity data that does not use the drag coefficient explicitly is sometimes used. The dimensionless terminal velocity is plotted against the dimensionless particle size as shown in Figure 3.11. This graph can be plotted for any of the models that have been described for the drag coefficient as well as for the experimental data. The graph shows the relationship between the two dimensionless variables explicitly and is the graphical equivalent of the Concha–Almendra analytical solution of the Abraham equation. The graphical representation does not require an analytical solution and it can be constructed purely numerically. This graph is particularly useful when both the particle size and the terminal settling velocity of a particle are known and an estimate of the sphericity of the particle is required. The reader is referred to the FLUIDS

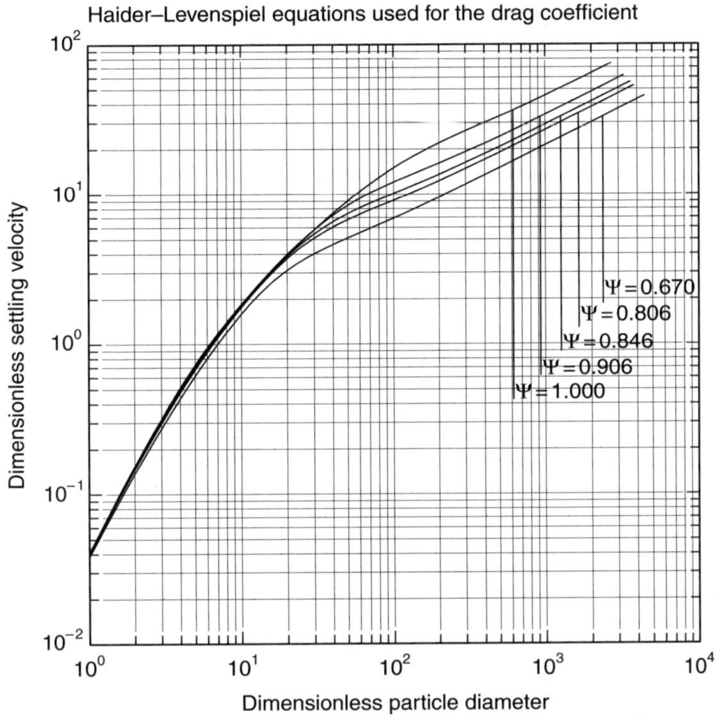

Figure 3.11 Generalized plot of dimensionless terminal settling velocity against the dimensionless particle size. Haider–Levenspiel equation used for the drag coefficient

computational toolbox to find this graph for each of the drag coefficient models.

3.4 Symbols used in this chapter

A_c Cross-sectional area of particles in plane perpendicular to direction of relative motion m^2.
a_p Surface area of particle m^2.
C_D Drag coefficient.
d_e Volume equivalent particle diameter m.
d_p Particle size m.
d_p^* Dimensionless particle size.
F_D Drag force N.
Re_p Particle Reynolds number.
V Relative velocity between particle and fluid m/s.
ν_p Volume of particle m^3.
V_T^* Dimensionless terminal settling velocity.
μ_f Viscosity of fluid Pa s.
ρ_f Density of fluid kg/m^3.

ρ_s Density of solid kg/m^3.
Φ_1 $C_D Re_p^2$.
Φ_2 Re_p/C_D.
ψ Sphericity.
Superscripts
* Indicates that variable is evaluated at the terminal settling velocity.
Subscripts
M Indicates modified value to take account of non-spherical shapes.

3.5 Practice problems

1. Calculate the terminal settling velocity of a 12-mm PMMA sphere of density 1200 kg/m^3 in water. Do the calculation manually using the Concha–Almendra method and also using the Karamanev equation and then compare the answers against the result from each method that is available in the FLUIDS toolbox.
2. A PMMA sphere having density 1200 kg/m^3 was found to have a terminal settling velocity of 0.242 m/s in water. Calculate the diameter of the particle. Do the calculation manually using the Concha–Almendra method and using equation 3.8 and then compare the answers against the result from each method that is available in the FLUIDS toolbox.
3. The terminal settling velocity of a plastic sphere of diameter 6.2 mm was measured to be 6.5 cm/s in water. Calculate the density of the material from which the sphere was made.
   Density of water = 1000 kg/m^3.
   Viscosity of water = 0.001 kg/ms.
   Use the Abraham equation.
4. Calculate the terminal settling velocities for the following particles in water at 25 °C
   3-mm glass sphere of density 2820 kg/m^3.
   12-mm PMMA sphere.
   0.1-mm stainless steel sphere of density 7800 kg/m^3.
   9.4-mm ceramic sphere of density 3780 kg/m^3.
5. Calculate the particle Reynolds number and the drag coefficient at terminal settling velocity for a 0.5-mm diameter glass sphere.
6. The terminal settling velocity for a limestone particle was measured to be 0.52 m/s in water at 25 °C. The density of limestone is 2750 kg/m^3 and the particle weighed 1.43 g. Calculate the equivalent volume diameter of the particle. Calculate the sphericity ψ of the particle. Calculate the modified and actual drag coefficient and the modified and actual Reynolds number at terminal settling velocity.
7. A dime is a disc approximately 17.8 mm in diameter and 1.25 mm thick and it weighs 2.31 g. The terminal settling velocity was measured in water to be 0.327 m/s. Calculate the drag coefficient at terminal settling velocity of the dime. If you do not know which dimension the dime will present to

the water when settling, determine this by a simple experiment. Explain why the dime adopts this attitude.

8. Calculate the volume, surface area and cross-sectional area perpendicular to the direction of motion of the following particles.
   A solid cube of side $20 \times 20 \times 40$ mm.
   A disk of diameter 17.8 mm and thickness 1.25 mm.
9. What is the terminal settling velocity of a 150 μm diameter spherical particle of density 3145 kg/m^3 settling in water ($\rho = 1000$ kg/m^3, $\mu = 0.001$ Pa s) and in air ($\rho = 1.2$ kg/m^3, $\mu = 17.5 \times 10^{-6}$ Pa s)?
10. What is the terminal settling velocity of the particle of the previous example settling in water in a 0.5 m radius centrifuge that rotates at 2000 rpm?
11. If Stokes' law is valid whenever $Re_p \leq 0.2$, calculate the largest diameter alumina sphere that can be modeled using Stokes' law at terminal settling conditions in water. The density of alumina is 2700 kg/m^3.
12. The FLUIDS toolbox provides you with convenient tools to calculate terminal settling velocities for all of the theoretical models that are discussed in the text. Not surprisingly these methods all give different answers. Since the toolbox makes it equally easy to use any of the methods you will need to formulate a strategy for deciding which method to use in any particular circumstance. Consider the following situations:

    (a) You want a quick calculated value of the terminal settling velocity of a 1-mm glass sphere in water.
    (b) You want a quick calculated value for the size of a sphere that has a terminal settling velocity of 10 cm/s in water.
    (c) You want an estimate of the sphericity of broken quartz particles from measurements of the terminal settling velocities.
    (d) When you calculate the terminal settling velocity of a particle you notice that $Re_p > 2 \times 10^3$.
    (e) You want to embed the calculation in a spreadsheet to analyze experimental data.
    (f) You want to embed the calculation in a C++ program to analyze data using the correlations for pressure drop in a slurry pipeline using the methods that are discussed in Chapter 4.
    (g) Your computer runs under the Unix operating system.
    (h) You are asked to give a talk to the History of Technology group at your local high school and you decide to say something about the influence of Fluid Mechanics in engineering during the twentieth Century. You decide to measure terminal settling velocities of some simple particles to illustrate your talk and you plan to show your audience what it was like to make the calculation when a slide rule was the only available computational tool.

Bibliography

The literature dealing with the drag coefficient of particles is large. Many empirical expressions for the drag coefficient have been presented. Clift *et al.* (1978) attempted to fit the available data using a set of equations each of which is valid over a restricted range of particle Reynolds number. Although this method produces a good fit to the data, the method is clumsy and the lack of continuity between the fitting equations at the ends of each range can lead to computational difficulties in some cases. Later authors (Turton and Levenspiel (1986), have shown that simpler equations provide superior fits at least to subsets of the available data and can be used to describe the drag coefficient of non-spherical particles also. There are many sets of data in the literature that have been determined and published over many years. The points shown in Figures 3.2, 3.3 and 3.4 are not actual data but averages from several investigators that were calculated and published by Lapple and Shepherd (1940). Several authors have presented empirical correlations between V_T^* and d_p^* but there does not seem to be any advantage over the use of the drag coefficient vs Φ_1 and Φ_2 that is used here and these results are not used in this book. Chhabra *et al.* (1999) have compared methods that are useful for non-spherical particles against about 1900 data points from the literature. They note that average errors in the calculated values of C_D in the range from 15 per cent to 25 per cent can be expected when using the correlations.

The use of stereological methods to measure the geometrical properties of irregularly shaped particles is described by Weibel (1980).

The importance of the terminal settling velocity in particle separation technology is discussed in King (2001).

References

Chhabra, R.P., Agarwal, L. and Sinha, N.K. (1999). Drag on non-spherical particles: an evaluation of available methods. *Powder Technology* **101**, 288–295.

Clift, R., Grace, J. and Weber, M.E. (1978). *Bubbles, Drops and Particles*. Academic Press.

Concha, F. and Almendra, E.R. (1979). Settling velocities of particulate systems. *International Journal of Mineral Processing* **5**, 349–367.

Concha, F. and Barrientos, A. (1986). Settling velocities of particulate systems. Part 4 Settling of non-spherical isometric particles of arbitrary shape. *International Journal of Mineral Processing* **18**, 297–308.

Ganser, G.H. (1993). A rational approach to drag prediction of spherical and non-spherical particles. *Powder Technology* **77**, 143–152.

Haider, A. and Levenspiel, O. (1989). Drag coefficient and terminal settling velocity of spherical and nonspherical particles. *Powder Technology* **58**, 63–706.

Karamanev, D.G. (1996). Equations for the calculation of the terminal velocity and drag coefficient of solid spheres and gas bubbles. *Chemical Engineering Communications* **147**, 75–84.

King, R.P. (2001). *Modeling and Simulation of Mineral Processing Systems*. Butterworth-Heinemann.

Lapple, C.E. and Shepherd, C.B. (1940). Calculation of particle trajectories. *Industrial and Engineering Chemistry* **32**, 605.

Pettyjohn, E.S. and Christiansen, E.B. (1948). Effect of particle shape on free settling rates of isometric particles. *Chemical Engineering Progress* **44**, 159–172.

Turton, R. and Levenspiel, O. (1986). A short note on the drag correlation for spheres. *Powder Technology* **47**, 83–86.

Weibel, E.R. (1980). *Stereological Methods. Volume 2, Theoretical Foundations.* John Wiley and Sons.

4

Transportation of slurries

The most important application of fluid flow techniques in the mineral processing industry is the transportation of slurries. Whenever solid materials are in particulate form transportation in the form of a slurry is possible. When the carrier fluid is water the method is referred to as hydraulic transportation and when the carrier fluid is air, pneumatic transportation.

There are two broad classifications for hydraulic transportation depending on whether the particles in the slurry can settle under the influence of the gravitational field or whether they are held more or less permanently in the suspension because of the rheological properties of the slurry itself. Slurries in these two classes are referred to as settling or heterogeneous and non-settling or homogeneous respectively. Non-settling slurries usually exhibit non-Newtonian behavior while settling slurries reflect the rheological properties of the pure carrier fluid.

4.1 Flow of settling slurries in horizontal pipelines

When a settling slurry is transported significant gradients in the solids concentration develop under the influence of gravity. The solid particles that are present in the slurry generate additional momentum transfer processes that must be considered when developing models for the transfer of momentum from the slurry to the pipe wall. The presence of solid particles increases the rate at which momentum is transferred between the fluid and the containing walls of the conduit. The transported particles frequently strike the walls and in so doing transfer momentum to the wall and dissipate some of their kinetic energy. The particles also transfer some of their momentum to the fluid if they are moving faster than the fluid in their neighborhood and receive momentum from the fluid when moving slower than the fluid in their neighborhood. These processes ensure a continuous exchange of momentum between the fluid and the walls, between the fluid and the particles and between the particles and the wall. This is illustrated in Figure 4.1.

The net result of this model is the existence of an additional path through which momentum can be transferred from the fluid to the solid wall and that is the indirect path from fluid to particles and from particles to the wall. This path acts in parallel with the direct transfer path from the fluid to the walls. This additional transfer mechanism leads to an increase in the pressure drop

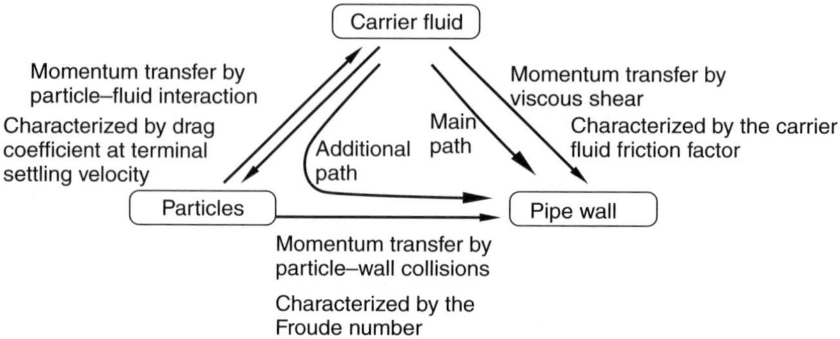

Figure 4.1 The additional paths for the transfer of momentum between the fluid and the wall when a slurry flows through a pipe

which is added to the pressure drop due to the carrier fluid alone. This idea is modeled by the concept of additive pressure drop.

$$\Delta P_{\mathrm{f,sl}} = \Delta P_{\mathrm{fw}} + \text{additional pressure drop due to particles} \tag{4.1}$$

This is expressed quantitatively in terms of the dimensionless group

$$\Phi = \frac{\Delta P_{\mathrm{f,sl}} - \Delta P_{\mathrm{fw}}}{\Delta P_{\mathrm{fw}}} \tag{4.2}$$

which represents the fractional increase in pressure gradient over and above that produced by the carrier fluid if it were flowing without particles at the same velocity as the slurry. $\Delta P_{\mathrm{f,sl}}$ is the pressure drop due to friction with the slurry flowing in the channel and ΔP_{fw} is the pressure gradient due to friction if the carrier fluid were flowing alone at the same velocity as the slurry. ΔP_{fw} can be calculated using the methods described in Chapter 2. Using equation 2.5

$$-\Delta P_{\mathrm{fw}} = 2 f_{\mathrm{w}} \rho_{\mathrm{w}} \bar{V}^2 \frac{L}{D} \tag{4.3}$$

A friction factor for the slurry can be defined analogously to equation 2.5

$$-\Delta P_{\mathrm{f,sl}} = 2 f_{\mathrm{sl}} \rho_{\mathrm{w}} \bar{V}^2 \frac{L}{D} \tag{4.4}$$

where ρ_{w} is the density of the carrier fluid and *not* the density of the slurry. Then Φ can be written in terms of the friction factors

$$\Phi = \frac{f_{\mathrm{sl}} - f_{\mathrm{w}}}{f_{\mathrm{w}}} \tag{4.5}$$

where f_{sl} is the friction factor for the slurry and f_{w} is the friction factor for the carrier fluid flowing at the same velocity as the slurry in the channel. Note that some authors prefer to use the density of the slurry rather than the density of water in equation 4.4. The values of the friction factor f_{sl} are somewhat different under these two conventions.

The frictional dissipation of energy is given by equation 2.7

$$F = \frac{-\Delta P_{f,sl}}{\rho_{sl}} = 2 f_{sl} \bar{V}^2 \frac{L}{D} \frac{\rho_w}{\rho_{sl}} \text{ J/kg of slurry} \qquad (4.6)$$

The value of F given by equation 4.6 can be used in the mechanical energy balance equation 2.40.

4.2 Four regimes of flow for settling slurries

The tendency that the solid particles have to settle under the influence of gravity has a significant effect on the behavior of a slurry that is transported in a horizontal pipeline. The settling tendency leads to a significant gradation in the concentration of solids in the slurry. The concentration of solids is higher in the lower sections of the horizontal pipe. The extent of the accumulation of solids in the lower section depends strongly on the velocity of the slurry in the pipeline. The higher the velocity the higher the turbulence level and the greater the ability of the carrier-fluid to keep the particles in suspension. It is the upward motion of eddy currents transverse to the main direction of flow of the slurry that is responsible for maintaining the particles in suspension. At very high turbulence levels the suspension is almost homogeneous with very good dispersion of the solids while at low turbulence levels the particles settle towards the floor of the channel and can in fact remain in contact with the flow and are transported as a sliding bed under the influence of the pressure gradient in the fluid. Between these two extremes of behavior, two other more-or-less clearly defined flow regimes can be identified. When the turbulence level is not high enough to maintain a homogenous suspense but is still sufficiently high to prevent any deposition of particles on the floor of the channel, the flow regime is described as being heterogeneous suspension. As the velocity of the slurry is reduced further a distinct mode of transport known as saltation develops. In the saltation regime, there is a visible layer of particles on the floor of the channel and these are being continually picked up by turbulent eddies and dropped to the floor again further down the pipeline. The solids therefore spend some of their time on the floor and the rest in suspension in the flowing fluid. Under saltation conditions the concentration of solids is strongly non-uniform. The flow regime depends strongly on the size and density of the particles that make up the slurry. For example, a higher level of turbulence is required to keep larger and heavier particles in suspension than is required for smaller and less dense particles. The four regimes of flow are illustrated in Figure 4.2.

The relationship between frictional pressure gradient and the slurry velocity varies from regime to regime and they can be delineated approximately in the particle size – slurry velocity space as shown in Figure 4.3. Small particle in a fast-moving slurry will be dispersed fairly uniformly through

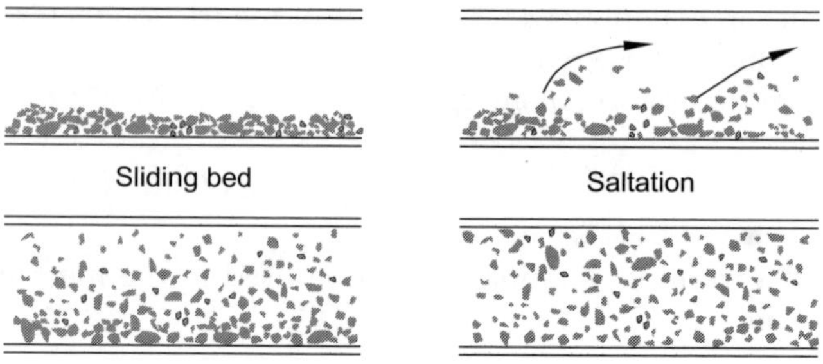

Figure 4.2 The four regimes of flow for settling slurries in horizontal pipelines

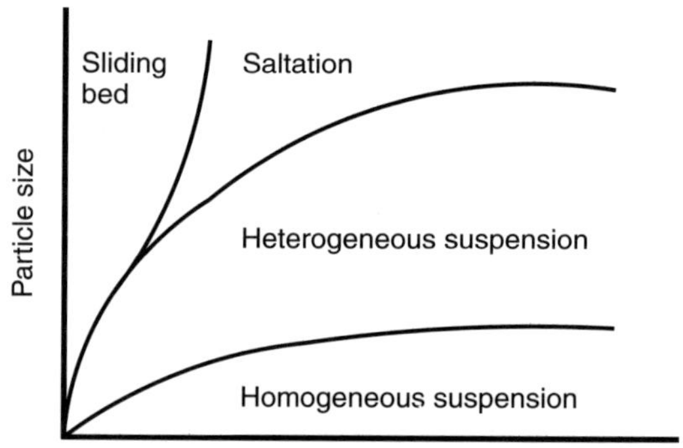

Figure 4.3 Schematic representation of the boundaries between the flow regimes for settling slurries in horizontal pipelines

the slurry while larger particles will have a greater tendency to settle and will produce a heterogeneous dispersion. Notice that the saltation regime pinches out at low velocities leaving only the three other regimes.

4.2.1 Saltation and heterogeneous suspension

The saltation and heterogeneous suspension regimes have been studied most widely and the best known correlation for the excess pressure gradient due to the presence of solid particles in the slurry is due to Durand, Condolios and Worster.

The pipe Froude number is a dimensionless group that indicates the relative strengths of the suspension and settling tendencies of the particles in the slurry.

$$Fr = \frac{\bar{V}^2}{gD(s-1)} \tag{4.7}$$

where s is the specific gravity of the solid and D is the diameter of the pipe.

The Froude number is a useful index for the importance of the momentum transfer process from particles to the pipe wall relative to the direct transfer of momentum from the fluid to the pipe wall. This is depicted as the lower horizontal arrow in Figure 4.1. Lower values of *Fr* indicate stronger particle–wall interactions relative to the fluid–wall interaction. Consequently, the fractional increase in pressure drop, Φ, varies inversely with *Fr*. The interaction between the fluid and the particles (the left-hand edge of the triangle in Figure 4.1) is summarized by the drag coefficient at terminal settling velocity C_D^*. This can be rationalized by noting that the relative velocity between particles and fluid originates with the difference of density between the fluid and the solid and the consequent settling of the solids relative to the fluid under the influence of gravity. Although there will always be a wide range of relative velocities between individual particles and the turbulent fluid in the pipe at any instant, C_D^* is used as the average value of C_D that describes the totality of fluid–particle interactions that occur. Although these arguments are only approximate and suggestive, they have led to some useful correlations for Φ in terms of the operating characteristics of real slurry flows.

The Durand–Condolios–Worster correlation for the excess pressure gradient is

$$\Phi = \frac{\Delta P_{f,sl} - \Delta P_{fw}}{\Delta P_{fw}} = \Omega C(\sqrt{C_D^*} Fr)^{-1.5} \tag{4.8}$$

where Ω is a constant and C is the volumetric fraction of solids in the suspension and C_D^* is the drag coefficient at terminal settling velocity. The value to be used for the constant Ω is uncertain and values between 65 and 150 are reported in the literature. Because this correlation does not apply to all regimes of flow, the experimental data cannot be used to fix the value more precisely. Errors of 100 per cent and more in the calculated value of Φ can result. This is not as serious as it might appear at first sight since in many cases the excess pressure drop is only a small fraction of the total pressure drop along the pipe and errors in the value of Φ are correspondingly less important. The pressure drop due to friction is sometimes specified as head loss per unit length of pipe. The head loss can be specified in terms of head of water or head of slurry.

$$i = -\frac{\Delta P_{f,sl}}{g \rho_w L} = 2 f_{sl}(s-1) Fr \text{ m water/m of pipe length} \tag{4.9}$$

or

$$j = -\frac{\Delta P_{f,sl}}{g \rho_{sl} L} = 2 f_{sl} \frac{\rho_s - \rho_w}{\rho_{sl}} Fr \text{ m slurry/m pipe length} \tag{4.10}$$

using equations 4.4 and 4.7.

Illustrative example 4.1

Use the Durand–Condolios–Worster correlation to calculate the pressure gradient due to friction when a slurry made from 1-mm silica particles is pumped through a horizontal 5-cm diameter pipeline at 3.5 m/s. The slurry contains 30 per cent silica by volume. The density of silica is $2700\,\text{kg/m}^3$, $\rho_\text{w} = 1000\,\text{kg/m}^3$, $\mu_\text{w} = 0.001\,\text{kg/ms}$. Use a value of 82 for Ω.

Solution

Use the toolbox to get the drag coefficient at terminal settling velocity as shown in Figure 4.4.

$$C_\text{D}^* = 0.945$$

$$Fr = \frac{\bar{V}^2}{g(s-1)D} = \frac{3.5^2}{9.81(2.7-1)0.05} = 14.69$$

$$\Phi = 82\ C\left(\sqrt{C_\text{D}^*}\ Fr\right)^{-1.5}$$

$$= \frac{82 \times 0.3}{(\sqrt{0.945}\ 14.69)^{1.5}} = 0.456$$

$$\frac{\Delta p_\text{f,sl}}{L} = \frac{\Delta p_\text{fw}}{L}(1+\Phi)$$

$$= \frac{2f_\text{w}\rho_\text{w}\bar{V}^2}{D}(1+0.456)$$

Use the toolbox to get the value of the friction factor f_w as shown in Figure 4.5.

$$Re = \frac{D\bar{V}\rho_\text{w}}{\mu_\text{w}} = \frac{0.05 \times 3.5 \times 1000}{0.001} = 1.75 \times 10^5$$

$$f_\text{w} = 0.00389$$

$$\frac{\Delta p_\text{f,sl}}{L} = \frac{2 \times 0.00389 \times 1000 \times 3.5^2 \times 1.456}{0.05} = 2.78\,\text{kPa/m}$$

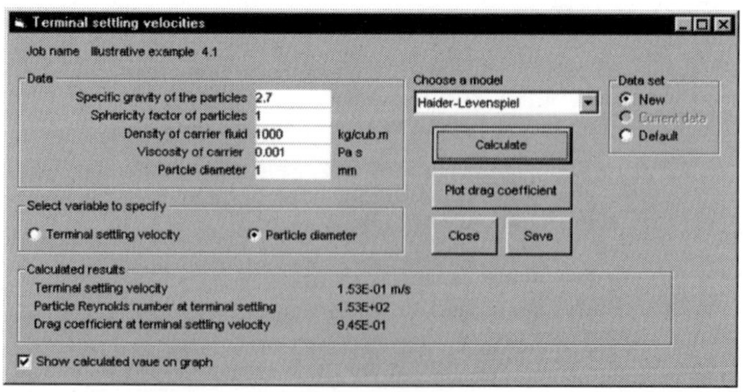

Figure 4.4 Data input screen to calculate the drag coefficient at terminal settling velocity fluid

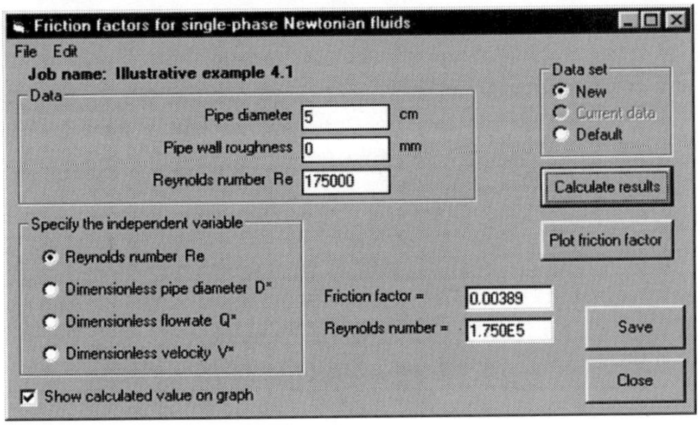

Figure 4.5 Data input screen to calculate the friction factor for the carrier fluid

4.2.2 Velocity at minimum pressure drop

One important consequence of the additional momentum transfer path is that the pressure drop in a pipe carrying slurry does not increase monotonically with the slurry velocity. There is a distinct velocity at which the pressure drop is a minimum. This can be seen in Figure 4.6 where the pressure gradient due to friction, calculated from equation 4.8, is plotted against the velocity for

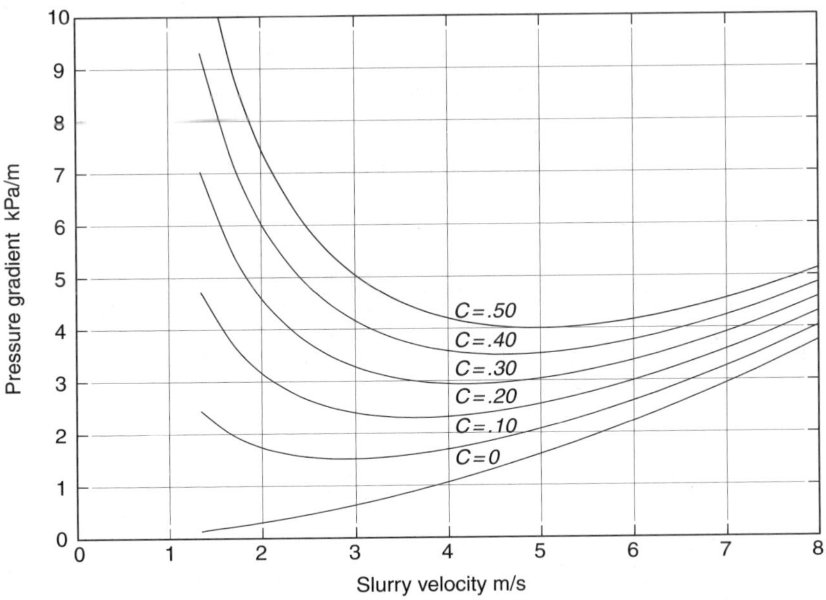

Figure 4.6 Frictional pressure gradient in a 10-cm pipe carrying a slurry of 1-mm silica particles. The volume fraction of solid in the slurry is *C*. The Durand–Condolios–Worster correlation was used with $\Omega = 82$ to generate the curves

slurries of different composition made from 1 mm quartz particles. The frictional pressure gradient that would result if only water were flowing in the pipe is shown as the curve with $C = 0$. As expected this curve shows an increasing pressure gradient as the velocity of the water increases but the curves for slurry show a clear minimum that occurs at higher velocities as the concentration of solid in the slurry increases.

The occurrence of this minimum in the pressure drop vs velocity curve has significance for slurry pipeline design. Clearly there is no merit in operating a pipeline at a velocity below the minimum because that would incur more energy loss at lower capacity. A rule of thumb that can be used is to choose the velocity to be approximately 20 per cent larger than the velocity at minimum pressure drop.

The velocity at minimum pressure drop is easy to calculate by finding the point on the curve that has zero slope.

$$\Delta P_{\mathrm{f,sl}} = 2 f_{\mathrm{w}} \rho_{\mathrm{w}} \bar{V}^2 \frac{L}{D} \left(1 + \Omega C \left(\frac{gD(s-1)}{\bar{V}^2 \sqrt{C_{\mathrm{D}}^*}} \right)^{1.5} \right) \tag{4.11}$$

$$\frac{\mathrm{d}\Delta P_{\mathrm{f,sl}}}{\mathrm{d}\bar{V}} = 2 f_{\mathrm{w}} \rho_{\mathrm{w}} \frac{L}{D} \left(2\bar{V} - \Omega C \bar{V}^{-2} \left(\frac{gD(s-1)}{\sqrt{C_{\mathrm{D}}^*}} \right)^{1.5} \right) = 0 \tag{4.12}$$

The variation of f_{w} with flowrate has been neglected in the differentiation. The velocity at minimum pressure drop is given by the solution to equation 4.12

$$\bar{V}_{\mathrm{opt}}^3 = \frac{\Omega}{2} C \left(\frac{gD(s-1)}{\sqrt{C_{\mathrm{D}}^*}} \right)^{1.5} \tag{4.13}$$

In practice it is more usual to choose the diameter of the pipe that will transport a given quantity of slurry at minimum pressure drop. In this case, the solution is slightly different and is obtained by substituting for $\bar{V}$ in terms of the pipe diameter before differentiating.

$$\bar{V} = \frac{4Q}{\pi D^2} \tag{4.14}$$

Setting the derivative with respect to D equal to 0 gives

$$D_{\mathrm{opt}}^{7.5} = \frac{128\, Q^3}{\pi^3\, \Omega C} \left(\frac{\sqrt{C_{\mathrm{D}}^*}}{g(s-1)} \right)^{1.5} \tag{4.15}$$

from which the optimum value of the pipe diameter can be determined once the volumetric flowrate and the properties of the particles are known.

Illustrative example 4.2

Calculate the diameter of the pipeline that is required to transport, at minimum pressure gradient, 120 tonnes/hr of silica sand as a slurry at 30 per cent solid by volume. Assume spherical part 1 mm diameter. Density of silica 2700 kg/m^3, $\rho_w = 1000$ kg/m^3, μ_w g/ms.

Solution

$$Q = \frac{120 \times 10^3}{3600 \times 2700} \quad 0.0412\,\text{m}^3/\text{s}$$

$$D_{opt}^{7.5} = \frac{128}{\pi^3}\frac{Q}{82C}\left(\frac{}{g}\right.$$

$$= \frac{128}{\pi^3} \times \frac{0.0}{8.2} \quad \left.\frac{\sqrt{0.812}}{9.81(2.7-1)}\right)^{1.5}$$

$$= 1.474 \times 1$$

$$D_{opt} = 0.123\,\text{m}$$

The average velocity in the pipel

$$\bar{V} = \quad = 3.5\,\text{m/s}$$

4.3 Head loss corr ions for separate flow regimes

While the Durand–Candoli rster correlation is useful in the heterogeneous suspension flow reg deviates more and more from actual conditions in the other regimes . Experimental observations have shown that different correlations shou used in each of the identifiable flow regimes. Although this is a logical approach it is not straightforward to apply. The main difficulty arises because it is not easy to define the boundaries between the flow regimes. These boundaries are poorly defined because they are based on visual observations of particle motions in small laboratory pipelines. Many researchers have attempted to establish correlations among the relevant experimental variables that can be used to define the boundaries of the flow regimes. These attempts have met with only limited success and an approach developed by Turian and Yuan (1977) is used here. This approach provides a completely self consistent definition of the flow regime boundaries that results directly from the head loss correlations and no additional correlations are required to define the boundaries. The method has the additional recommendation that it is based on a large data base of reliable experimental data and consequently the method can be used with confidence for practical engineering work.

Using the experimental data, Turian and Yuan established that the excess pressure gradient in each flow regime can be correlated using an equation of the form

$$f_{sl} - f_w = KC^{\alpha} f_w^{\beta} C_D^{*\,\gamma} Fr^{\delta} \tag{4.16}$$

The coefficients K, α, β, γ and δ have values that are specific to each flow regime. Using experimental data gathered from experiments in each flow regime, the best available values of these parameters in each flow regime are given by
Sliding bed (Regime 0)

$$f_{\mathrm{sl}} - f_{\mathrm{w}} = 12.13\, C^{0.7389} f_{\mathrm{w}}^{0.7717}\, C_{\mathrm{D}}^{*\,-0.4054}\, Fr^{-1.096} \tag{4.17}$$

Saltation (Regime 1)

$$f_{\mathrm{sl}} - f_{\mathrm{w}} = 107.1\, C^{1.018} f_{\mathrm{w}}^{1.046}\, C_{\mathrm{D}}^{*\,-0.4213}\, Fr^{-1.354} \tag{4.18}$$

Heterogeneous suspension (Regime 2)

$$f_{\mathrm{sl}} - f_{\mathrm{w}} = 30.11\, C^{0.868} f_{\mathrm{w}}^{1.200}\, C_{\mathrm{D}}^{*\,-0.1677}\, Fr^{-0.6938} \tag{4.19}$$

Homogeneous suspension (Regime 3)

$$f_{\mathrm{sl}} - f_{\mathrm{w}} = 8.538\, C^{0.5024} f_{\mathrm{w}}^{1.428}\, C_{\mathrm{D}}^{*\,0.1516}\, Fr^{-0.3531} \tag{4.20}$$

Turian and Yuan designate regime 0 as stationary bed but do not make a distinction between the condition in which the bed remains stationary and does not slide and that in which the bed slides along the lower inner surface of the pipe wall under the influence of the pressure gradient. Both conditions are included in regime 0. The distinction between the stationary bed and the sliding bed is strongly emphasized in the stratified flow model that is discussed in Section 4.4.

Fairly consistent trends in the variation of the correlating parameters can be seen in the four correlations. The influence of the Froude number becomes less pronounced as the flow changes from the sliding bed regime through saltation and heterogeneous suspension to homogeneous suspension. This reflects the decreasing influence of the particle settling process on the momentum transfer, and hence frictional dissipation, as the suspension becomes more homogeneous. Notice that the exponent on the Froude number is always negative. The exponent on C_{D}^{*} increases as the flow changes from sliding bed to homogeneous suspension reflecting the greater tendency of high drag coefficient particles to pick up momentum from the fluid and then to transfer it to the wall. The exponent on f_{w} increases by a factor of 2 reflecting the increasing influence of the direct momentum transfer process from carrier fluid to the wall as increasingly homogeneous flow is maintained.

4.3.1 Flow regime boundaries

The boundaries of the flow regimes are defined in a self-consistent manner by noting that any two regimes are contiguous at their common boundary and therefore each of the two correlation equations must be satisfied simultaneously. For example, the boundary between the sliding bed regime (Regime 0) and the saltation regime (Regime 1) must lie along the solution locus of the equation

$$12.13\, C^{0.7389} f_{\mathrm{w}}^{0.7717}\, C_{\mathrm{D}}^{*\,-0.4054}\, Fr^{-1.096} = 107.1\, C^{1.018} f_{\mathrm{w}}^{1.046}\, C_{\mathrm{D}}^{*\,-0.4213}\, Fr^{-1.354} \tag{4.21}$$

which is simplified to

$$Fr = \frac{\bar{V}^2}{Dg[s-1]} = 4679\, C^{1.083} f_{\mathrm{w}}^{1.064}\, C_{\mathrm{D}}^{*\;-0.0616} \tag{4.22}$$

The regime transition number for transitions between regime 0 and regime 1 is defined by

$$R_{01} = \frac{Fr}{4679\, C^{1.083} f_{\mathrm{w}}^{1.064}\, C_{\mathrm{D}}^{*\;-0.0616}} \tag{4.23}$$

and this number must be unity on the boundary between these two regimes. The transition numbers for the other possible transitions are found in the same way and are given by

$$R_{02} = \frac{Fr}{0.1044\, C^{-0.3225} f_{\mathrm{w}}^{-1.065}\, C_{\mathrm{D}}^{*\;-0.5906}} \tag{4.24}$$

$$R_{12} = \frac{Fr}{6.8359\, C^{0.2263} f_{\mathrm{w}}^{-0.2334}\, C_{\mathrm{D}}^{*\;-0.3840}} \tag{4.25}$$

$$R_{13} = \frac{Fr}{12.522\, C^{0.5153} f_{\mathrm{w}}^{-0.3820}\, C_{\mathrm{D}}^{*\;-0.5724}} \tag{4.26}$$

$$R_{23} = \frac{Fr}{40.38\, C^{1.075} f_{\mathrm{w}}^{-0.6700}\, C_{\mathrm{D}}^{*\;-0.9375}} \tag{4.27}$$

$$R_{03} = \frac{Fr}{1.6038\, C^{0.3183} f_{\mathrm{w}}^{-0.8837}\, C_{\mathrm{D}}^{*\;-0.7496}} \tag{4.28}$$

These numbers define the boundaries between any two flow regimes a and b by the condition

$$R_{ab} = 1 \tag{4.29}$$

It is usual to define the regime boundaries on a plot of particle size against slurry velocity. Equation 4.29 defines an implicit relationship between these two variables provided that the properties of the fluid and the volumetric concentration of solid in the slurry are specific. The diameter of the pipe must also be specified. The calculation is not easy because f_{w} is a complex function of the slurry velocity through equation 2.15 or Figure 2.2. In addition C_{D}^* is a function of the particle size which must be calculated using the method that are described in Chapter 3.

Leaving aside the problem of constructing the regime boundaries for the moment, it is possible to identify the regime that applies to a particular set of physical conditions quite simply from a knowledge of the transitions numbers R_{ab}. Consider the behavior of the R_{ab} numbers as the value of the velocity increases. Consider a slurry made up of particles of size d_{p} and specific gravity s in water. The drag coefficient at terminal setting velocity of these particles is fixed and can be calculated using any of the methods that were described in Chapter 3. If $a < b$ the value of R_{ab} increases monotonically as the

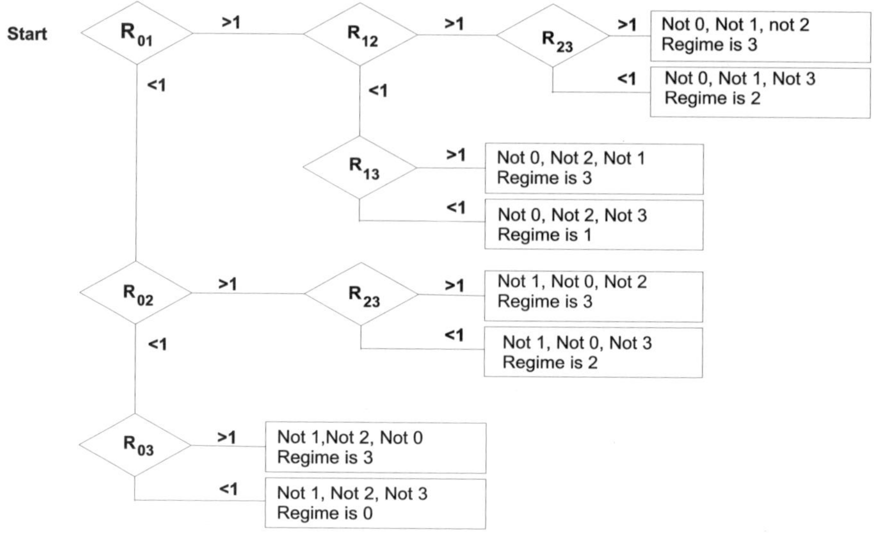

Figure 4.7 Decision tree for establishing flow regime

velocity increases. At low velocities, $R_{ab} < 1$ and with increasing velocity, the value of R_{ab} will eventually pass through the value 1.0. This must signal a transition out of regime a. The following simple rules will fix the flow regime at any combination of the variables $\bar{V}$ and d_p.

$$\text{If } R_{ab} < 1 \text{ the regime is not } b.$$
$$\text{If } R_{ab} > 1 \text{ the regime is not } a.$$

These inequalities must be tested for the combinations of ab as shown in the decision tree in Figure 4.7. No more than three of the transition numbers need be calculated to fix the flow regime uniquely. Notice that these rules test the flow regimes negatively and a single test will never suffice to define the flow regime. It is always necessary to test at least three different combinations of a and b to get a definitive identification of the flow regime.

The applicable flow regime can be identified quickly and easily using Figure 4.7 and the appropriate equation can be selected from equations 4.17, 4.18, 4.19 or 4.20 to calculate the slurry friction factor. The entire calculation, including the identification of the flow regime can be done most conveniently by selecting the Turian–Yuan correlation from the menu in the FLUIDS software toolbox.

Illustrative example 4.3

Use the Turian–Yuan correlation to calculate the pressure gradient due to friction when a slurry made from 1-mm silica particles is pumped through a horizontal 5-cm diameter pipeline at 3.5 m/s. The slurry contains 30 per cent silica by volume. $s = 2.7$, $\rho_w = 1000\,\text{kg/m}^3$, $\mu_w = 0.001\,\text{kg/ms}$.

Solution

The first step is to determine the flow regime.

Use the toolbox to get the drag coefficient at terminal settling velocity.

$$C_D^* = 0.815$$

Use the toolbox to get the friction factor for water.

$$Re = \frac{DV\rho_w}{\mu_w} = \frac{0.05 \times 3.5 \times 1000}{0.001} = 1.75 \times 10^5$$

$$f_w = 0.00389$$

$$Fr = \frac{\bar{V}^2}{g(s-1)D} = \frac{3.5^2}{9.81(2.7-1)0.05} = 14.69$$

Calculate the transition numbers.

$$\begin{aligned} R_{01} &= \frac{Fr}{4679\ C^{1.083} f_w^{1.064} C_D^{*-0.0616}} \\ &= \frac{14.69}{4679\ 0.3^{1.083}\ 0.00389^{1.064}\ 0.815^{-0.0616}} \\ &= 4.20 \end{aligned}$$

Using Figure 4.7 the next transition number to test is R_{12}

$$R_{12} = \frac{14.69}{6.8359 \times 0.3^{0.2263}\, 0.00389^{-0.2334}\, 0.815^{-0.3840}} = 0.714$$

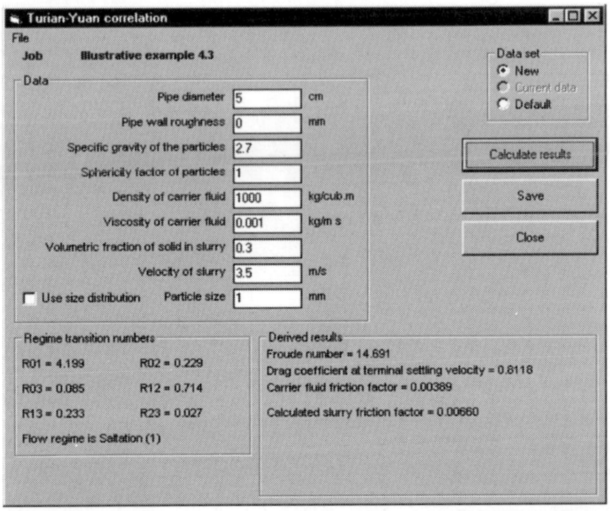

Figure 4.8 Data input and calculation screen to calculate the slurry friction factor using the Turian–Yuan correlations

Using Figure 4.7 again, the next transition number to test is R_{13} and since $R_{13} < 1$, the flow is established as being in regime 1 (saltation). The slurry friction factor is calculated using equation 4.18

$$\begin{aligned} f_{sl} &= f_w + 107.1 \times 0.3^{1.018} \times 0.00389^{1.046} \times 0.815^{-0.4213} \times 14.69^{-1.354} \\ &= 0.00389 + 0.00272 = 0.0066 \end{aligned}$$

These calculations are tedious and the FLUIDS toolbox provides a convenient alternative for evaluating the transitions numbers and the friction factor (see Figure 4.8).

$$\Delta P_{f,sl} = 2 f_{sl} \rho_w \bar{V}^2 \frac{L}{D}$$

$$\frac{\Delta P_{f,sl}}{L} = \frac{2 \times 0.0066 \times 1000 \times 35^2}{0.05} = 3.23\,\text{kPa/m}$$

Compare this result with that obtained in illustrative example 4.1, which is based on the Durand–Condolios–Worster correlation.

Sometimes it is necessary to have an accurate picture of the entire flow regime diagram for a given slurry in a particular pipeline. This type of plot is illustrated generically in Figure 4.9. The regime boundaries can be generated by plotting all the curves that represent solutions of the equations $R_{ab} = 1$ for every combination of a and b with $a < b$. This produces a series of intersecting lines in the space of the variables d_p and $\bar{V}$ as shown in Figure 4.9. However, not all of the resulting lines represent valid regime boundaries and the physically realizable boundaries must be selected. This can be done most simply by noting that the graph of $R_{ab} = 1$ can represent only the boundary between regime a and b. Thus

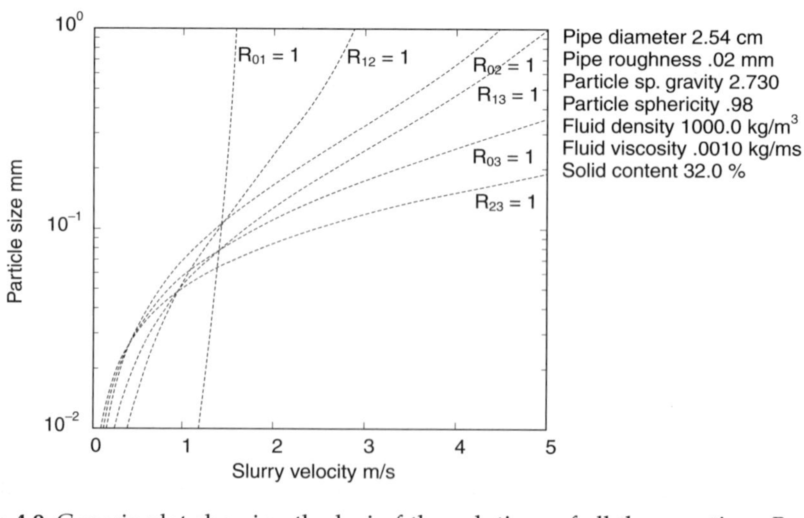

Figure 4.9 Generic plot showing the loci of the solutions of all the equations $R_{ab} = 1$.

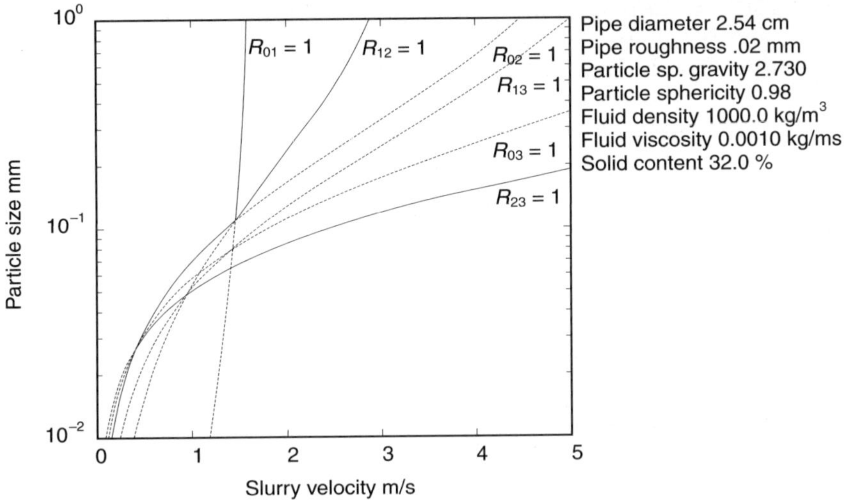

Figure 4.10 Real boundaries are selected using the exclusion rules

all non-physical boundary lines and non-physical segments of lines can be identified and eliminated. This selection process is illustrated in Figure 4.10 where all potential boundary lines are shown as dotted curves and the actual boundaries selected using the exclusion rules are shown as solid lines. The final regime boundary plot is shown in Figure 4.11. Remember that a completely new regime plot is required whenever any physical property of the slurry or particle

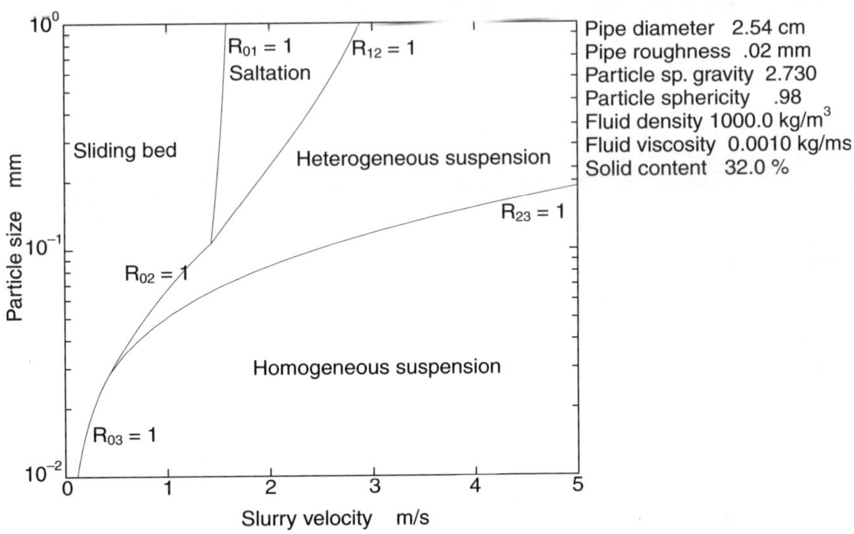

Figure 4.11 The flow regime diagram is constructed finally by eliminating all non-physical boundary lines

is changed. Every different pipe diameter also requires its own regime boundary plot. The construction of these regime diagrams requires a considerable amount of calculation. Consequently, they are not widely used in practice for engineering calculations. However, the diagrams can be generated quickly by choosing the Turian–Yuan regime plot item from the main menu of the FLUIDS software toolbox as shown in illustrative example 4.4.

Illustrative example 4.4

Display the flow regime plot for a 30 per cent by volume silica sand slurry that is pumped in a 12.3 cm diameter pipe.

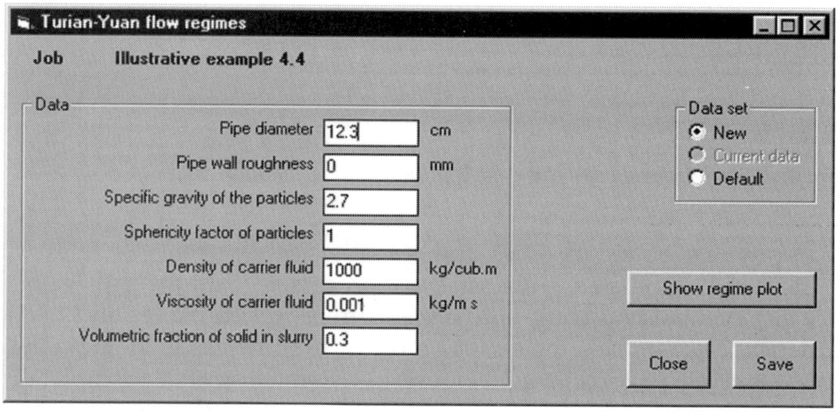

Figure 4.12 Data input screen that generates the Turian–Yuan regime boundaries

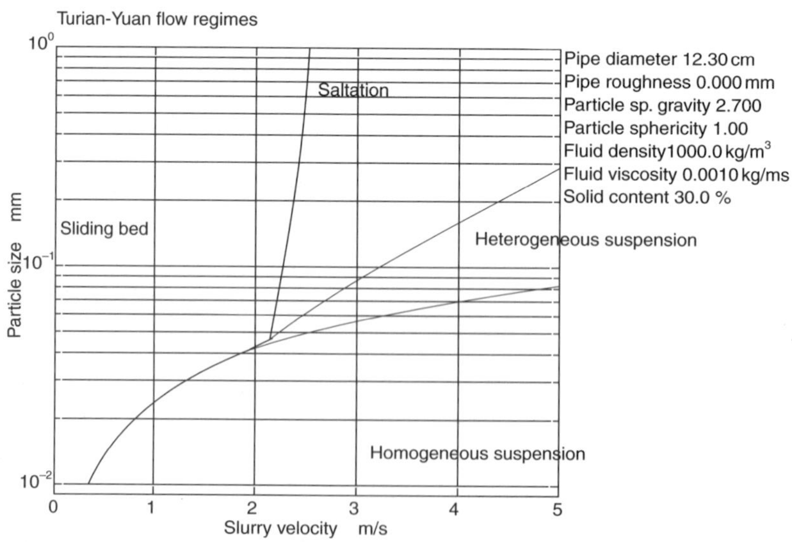

Figure 4.13 Turian–Yuan regimes for illustrative example 4.4

It is not practical to undertake this calculation manually, so use the FLUIDS toolbox (see Figure 4.12).

The inner surface of pipes that carry slurries wear smooth very quickly so the pipe wall roughness is taken to be zero.

Notice how the flow regime plot as shown in Figure 4.13 differs quite markedly from that shown in the Figure 4.11.

4.4 Head loss correlations based on a stratified flow model

Although the correlations that are described in the previous sections provide a self consistent approach to the calculation of the excess pressure gradient due to the presence of the solid particles, it is by no means certain that the correlations for the individual flow regimes are satisfactory under the full range of conditions that are of interest in industrial applications. The boundaries of the four flow regimes that are well defined in terms of the defining empirical equations for the relative excess pressure loss, do not have corresponding sharp transitions in real slurry pipelines. An alternative approach is based on a continuous transition from fully stratified flow at low velocities to fully suspended or heterogeneous flow at higher velocities. This approach is described in Section 4.4.1.

4.4.1 Fully stratified flow

When the particles in the suspension are comparatively large and the velocity not too large, most of the particles settle to the bottom of the pipe and are transported as a sliding bed. Most of the solids are supported on the bottom of the pipe as contact load. The movement of this bed is resisted by the mechanical friction between the particles in the bed that are up against the pipe wall and the pipe wall. The resisting force can be calculated as

$$F_{fr} = \mu_s F_N$$

where μ_s is the coefficient of friction between the bed and the wall and F_N is the normal force between bed and pipe wall integrated over the portion of the pipe wall that is in contact with the particle bed.

Two forces act to move the bed along the wall: the frictional drag caused by the carrier fluid moving above the bed and the body force that acts on the bed caused by the pressure gradient in the direction of flow. The pressure gradient acts on the cross-sectional area of the bed as shown in Figure 4.14. The steady-state motion of the sliding bed is defined by the balance of these forces that tend to move the bed and the frictional force which resists the sliding motion.

Fully stratified flow is not usually encountered in practice although it may be an advantageous mode of transport under some circumstances. An analysis of the sliding bed behavior provides information on an important design constraint called the limit of stationary deposition. This is the velocity below which the bed ceases to slide and is thus clearly a lower limit for the slurry

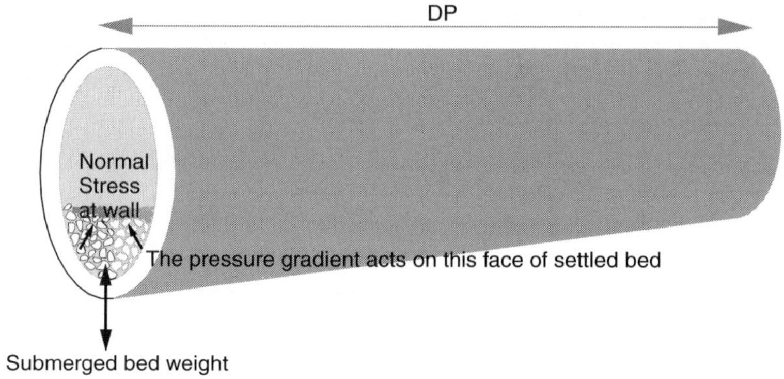

Figure 4.14 Fully stratified flow in a pipe

velocity if the solid is to be transported at all. This limiting velocity, which will be represented by V_s, varies with the solid content of the slurry. It is obviously zero for clear water and passes through a maximum V_{sm} at the critical deposit concentration C_{sm} before decreasing as the solid content increases towards its ultimate limit which is represented by the solid volume fraction of the loosely packed bed for the particles. V_{sm} can be calculated from (Wilson *et al.* 1997)

$$V_{sm} = 1.565 \frac{\left(\frac{D}{d_p}\right)^{0.7} d_p^{1.75}}{d_p^{1.3} + 1.1 \times 10^{-7} \left(\frac{D}{d_p}\right)^{0.7}} \left(\frac{s-1}{1.65}\right)^{0.55} \tag{4.30}$$

It is convenient to define the solid concentration relative to the solid volume fraction in a loosely packed bed of particles, C_{vb} so that

$$C_r = \frac{C}{C_{vb}} \tag{4.31}$$

Particles that are monosize and approximately spherical in shape have $C_{vb} = 0.6$.

The volumetric concentration of solids that has the maximum deposition velocity can be calculated from (Wilson *et al.* 1997)

$$C_r^* = 4.83 \times 10^{-4} \frac{D^{0.4}}{d_p^{0.84}} \left(\frac{1.65}{s-1}\right)^{0.17} \tag{4.32}$$

and the relative critical deposit concentration is given by

$$\begin{aligned} &= 0.66 \quad \text{if } C_r^* > 0.66 \\ C_{rm} &= C_r^* \quad \text{if } 0.05 \le C_r^* \le 0.66 \\ &= 0.05 \quad \text{if } C_r^* < 0.05 \end{aligned} \tag{4.33}$$

which reflect upper and lower limits that have been identified from experimental measurements.

The values of the maximum for the stationary deposit velocity, V_{sm}, and the critical deposit concentration, C_{sm}, provide a reference point from which the stationary deposition velocity can be calculated at any other volumetric concentration C using the equation

$$\frac{V_s}{V_{sm}} = 6.75\, C_r^{\alpha}\left(1 - C_r^{\alpha}\right)^2 \quad \text{with} \quad C_r = \frac{C}{C_{vb}} \tag{4.34}$$

and

$$\alpha = \frac{\ln(0.333)}{\ln(C_{rm})} \tag{4.35}$$

provided that $C_{rm} \leq 0.33$. If $C_{rm} > 0.33$

$$\frac{V_s}{V_{sm}} = 6.75(1 - C_r)^{2\beta}\left(1 - (1 - C_r)^{\beta}\right) \tag{4.36}$$

with

$$\beta = \frac{\ln(0.666)}{\ln(1 - C_{rm})} \tag{4.37}$$

The locus of stationary deposit velocities for two typical situations are shown in Figure 4.15 for large particles in a 30 cm pipe and in Figure 4.16 for smaller

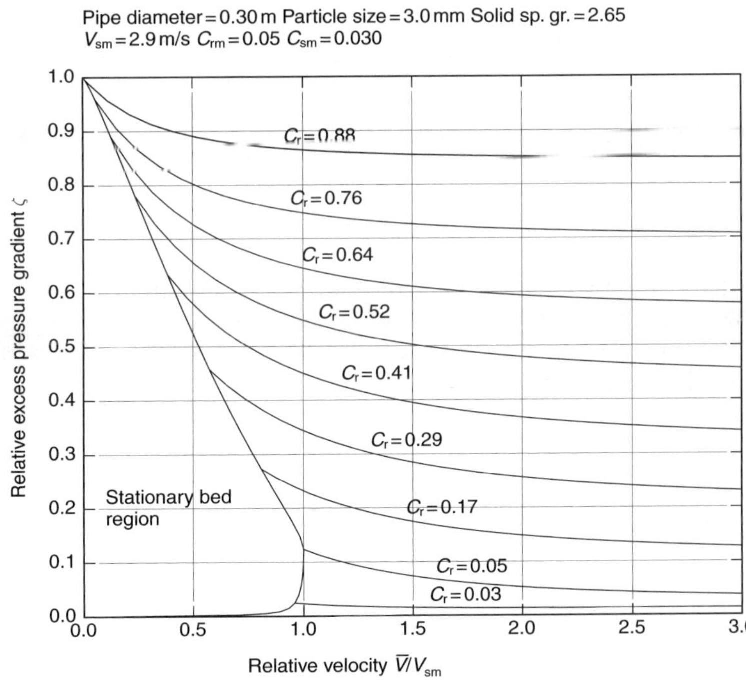

Figure 4.15 Excess pressure gradient for fully stratified flow of 3 mm sand slurry in a 30 cm pipe

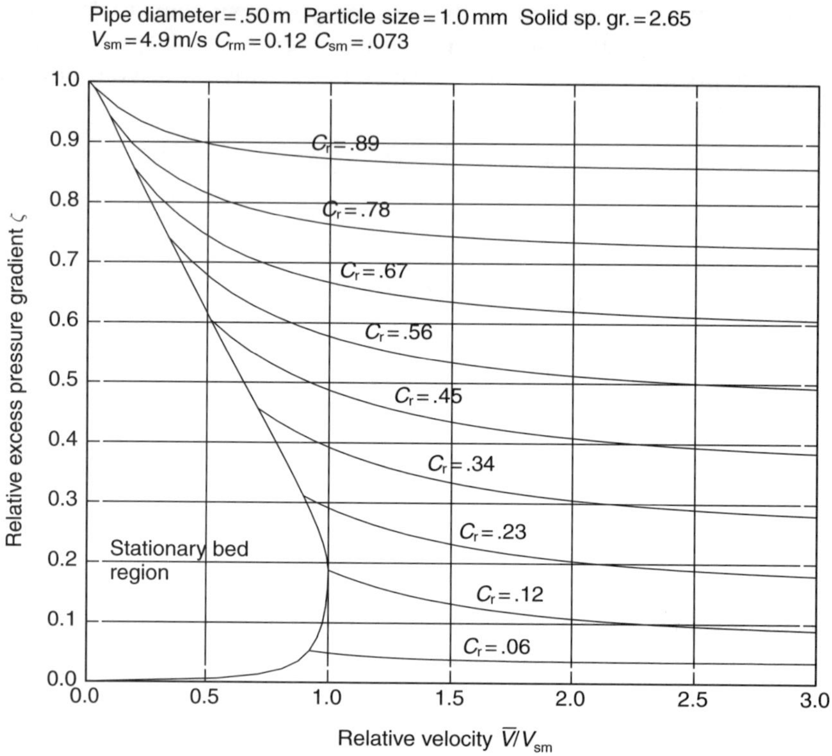

Figure 4.16 Excess pressure gradient for fully stratified flow of 1-mm coal slurry in a 50 cm pipe

particles in a larger diameter pipe. When the average velocity in the pipe exceeds the stationary deposit velocity V_s the bed slides and fully stratified flow results.

The excess pressure drop due to the energy dissipation between the solids in the bed and the pipe wall can be conveniently related to the pressure gradient that would be measured in a horizontal pipe filled with slurry at the loosely packed concentration C_{vb}. Such a slurry would flow as a plug and the pressure gradient would be entirely due to the solid friction between the bed of solids and the pipe wall. It can be shown that, under these conditions, the total normal force F_N exerted by the particles on the pipe wall is equal to twice the submerged weight of the particles. The hydraulic gradient for this plug flow condition is given by

$$i_{pg} = 2\,\mu_s(s-1)\,C_{vb} \quad \text{m water/m pipe} \tag{4.38}$$

The relative excess pressure gradient is defined by

$$\zeta = \frac{i_{sl} - i_w}{i_{pg}} \tag{4.39}$$

and ζ can be calculated at any slurry velocity from

$$\zeta = \zeta_\infty + \frac{1 - \zeta_\infty}{(1 + V_r)^a} \tag{4.40}$$

where

$$V_r = \frac{V}{V_{sm}} \tag{4.41}$$

and

$$\begin{aligned} a &= 3.6 - 5.2\,C_r(1 - C_r) && \text{for } C_r \geq C_{rm} \\ &= (3.6 - 5.2\,C_{rm}(1 - C_{rm}))\frac{C_{rm}}{C_r} && \text{for } C_r < C_{rm} \end{aligned} \tag{4.42}$$

The asymptotic limit of ζ at high velocities ζ_∞ is given by

$$\zeta_\infty = 0.5\,C_r(1 + C_r^{0.66}) \tag{4.43}$$

The relative excess pressure gradient is shown as a function of V_r for two typical cases in Figures 4.15 and 4.16.

The PGDTF for fully stratified flow can be conveniently calculated from these relationships

$$i_{sl} = i_w + \zeta\, i_{pg} \tag{4.44}$$

The application of this method is demonstrated in illustrative example 4.6.

Illustrative example 4.6 Fully stratified flow

This example is based on Case Study 5.2 in Wilson *et al.* (1997).

When dredging cohesive clays, the cutter head shaves off slices of clay which emerge from the pump discharge as fist-sized lumps, roughly spherical in shape, and about 100 mm in diameter. These lumps are transported as a slurry through a steel pipeline of 0.70 m internal diameter from the dredger to a disposal area. Calculate the pressure gradient in the horizontal pipeline and the energy dissipation due to friction.

Data:

Lump size = 100 mm
Lump density = 1790 kg/m^3
Volume fraction in loose packing = 0.6
Coefficient of sliding friction = 0.31
Carrier fluid density = 1020 kg/m^3 (sea water)
Carrier fluid viscosity = 0.001 Pa s
Surface roughness of pipe wall = 0.7 mm
Pumping rate = 1.77 m^3/s
Delivered solid volume fraction = 0.0714

The large size of the particles that are transported indicates that the flow will be fully stratified.

Calculate the critical concentration and velocity:

$$S_s = 1.79, \quad S_f = 1.02$$

$$C_r^* = \frac{4.83 \times 10^{-4} \times (0.7)^{0.4}}{(0.1)^{0.84}} \times \left(\frac{1.65}{0.75}\right)^{0.17} = 0.0033$$

Because $C_r^* < 0.05$, C_{rm} takes its lower limiting value of 0.05.

$$V_{sm} = \frac{1.565\left(\frac{0.7}{0.1}\right)^{0.7}(0.1)^{1.75}}{(0.1)^{1.3} + 1.1 \times 10^{-7}\left(\frac{0.7}{0.1}\right)^{0.7}} \times \left(\frac{0.75}{1.65}\right)^{0.55}$$

$$= 1.41\,\text{m/s}$$

Compare this with the slurry velocity in the pipeline,

$$V = \frac{Q}{\frac{\pi}{4}D^2} = \frac{1.77}{\frac{\pi}{4}0.7^2}$$

$$= 4.6\,\text{m/s}$$

This is well above V_{sm} so bed of clay will move in the pipe.

The relative velocity is

$$V_r = \frac{4.6}{1.41} = 3.26$$

The relative excess pressure drop can now be calculated.

$$C_r = \frac{0.0714}{0.6} = 0.119$$

$$\zeta_\infty = 0.5 \times 0.119 \times (1 + 0.119^{0.66})$$

$$= 0.0741$$

$$q = 3.6 - 5.2 \times 0.119 \times (1 - 0.119) = 3.05$$

$$\zeta = 0.0741 + \frac{(1 - 0.0741)}{(1 + 3.26)^{3.05}}$$

$$= 0.085$$

The plug flow pressure gradient is

$$i_{pg} = 2 \times 0.31 \times 0.77 \times 0.6$$

$$= 0.285\,\text{m water/m}$$

These calculations are straightforward but tedious, and they can be avoided by using the stratified flow feature in the FLUIDS toolbox as shown in the

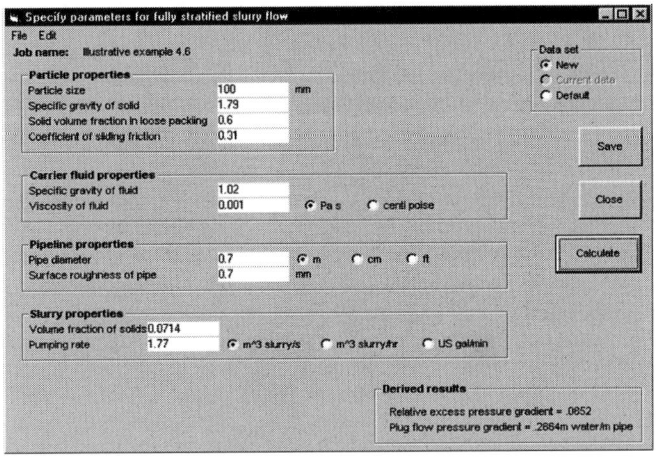

Figure 4.17 Data input screen for calculating the pressure drop under fully stratified flow

data input screen that is shown in Figure 4.17. The hydraulic gradient for water can be obtained from the friction factor feature of the FLUIDS toolbox.

$$
\begin{aligned}
Re &= \frac{0.7 \times 4.6 \times 1090}{0.001} \\
&= 2.38 \times 10^6
\end{aligned}
$$

The friction factor of water flowing in the pipe is $f_w = 0.00476$.

$$
i_w = \frac{2 \times 0.00465 \times 4.6^2}{0.7} = 0.288 \text{ m water/m}
$$

The pressure gradient when transporting the dredger output is

$$
i_{sl} = i_w + \zeta\, i_{pg} = 0.288 + 0.085 \times 0.286 = 0.312 \text{ m water/m}
$$

4.4.2 Heterogenous flow of settling slurries

Heterogenous flow conditions exist when most of the particles are supported by the fluid and the contact load between settled solids on the pipe wall is negligible. The particles are suspended only when the average velocity is high. At high fluid velocities the particles can be distributed more or less uniformly across the pipe cross-section and the slurry behaves as a homogenous fluid. Between the condition of fully stratified flow that was described in the previous section and homogeneous flow there is a more or less continuous transition through conditions when the slurry is heterogeneous or partially stratified with the particles concentrating in the lower regions of the pipe. The solid particles contribute a steadily decreasing fraction of a total energy dissipation as the slurry becomes more homogenous. In this section, an empirical method is described which can be used to calculate the pressure gradient due to friction and the frictional energy dissipation.

The method is based on the fractional increase in pressure gradient due to the presence of the solid

$$\Phi = \frac{\Delta P_{sl} - \Delta P_{fw}}{\Delta P_{fw}} \tag{4.45}$$

Φ decreases as the average velocity of the slurry in the pipe increases. Suspended solids are transported more efficiently than contact load and this causes the decrease in the excess frictional energy dissipation as velocity increases. An empirical correlation that has been proposed by Wilson *et al.* (1997) is

$$\Phi = \frac{C(s-1)\,\mu_s\, gD}{4 f_w\, \bar{V}^2}\left(\frac{\bar{V}_{50}}{\bar{V}}\right)^M \tag{4.46}$$

$\bar{V}_{50}$ is the velocity at the midpoint of the transition from fully stratified to homogenous flow. Like the Durand–Condolios–Worster correlation this method postulates that Φ is directly proportional to the volume fraction of solid in the slurry. The parameter M accounts to some extent for the distribution of particle sizes in the slurry and is calculated from the 50 per cent passing size d_{50} and the 85 per cent passing size d_{85} of the particle population. This relationship is quite complex and is given by

$$M = \min[(0.25 - 13\,\sigma_s^2)^{-1/2}, 1.7] \tag{4.47}$$

$$\sigma_s = \log_{10}\left(\frac{w_{85}\cosh(60\, d_{85}/D)}{w_{50}\cosh(60\, d_{50}/D)}\right) \tag{4.48}$$

w_{85} and w_{50} are related to the terminal settling velocities of the particles of size d_{85} and d_{50} respectively

$$w = 0.9\,\nu_T + 2.7\left(\frac{(\rho_s - \rho_f) g\, \mu_f}{\rho_f^2}\right)^{1/3} \tag{4.49}$$

The terminal settling velocities are calculated using any of the methods that are described in Chapter 3. w_{50} is calculated from equation 4.49 by substituting ν_T for particles of size d_{50} and w_{85} by substituting v_T for particles of size d_{85}.

$\bar{V}_{50}$ is a function of the terminal settling velocity of the d_{50} particle and the ratio d_{50}/D.

$$\bar{V}_{50} = w_{50}\left(\frac{2}{f_w}\right)^{1/2}\cosh\left(60\frac{d_{50}}{D}\right) \tag{4.50}$$

Two typical cases are shown in Figures 4.18 and 4.19.

The method is demonstrated in illustrative example 4.7.

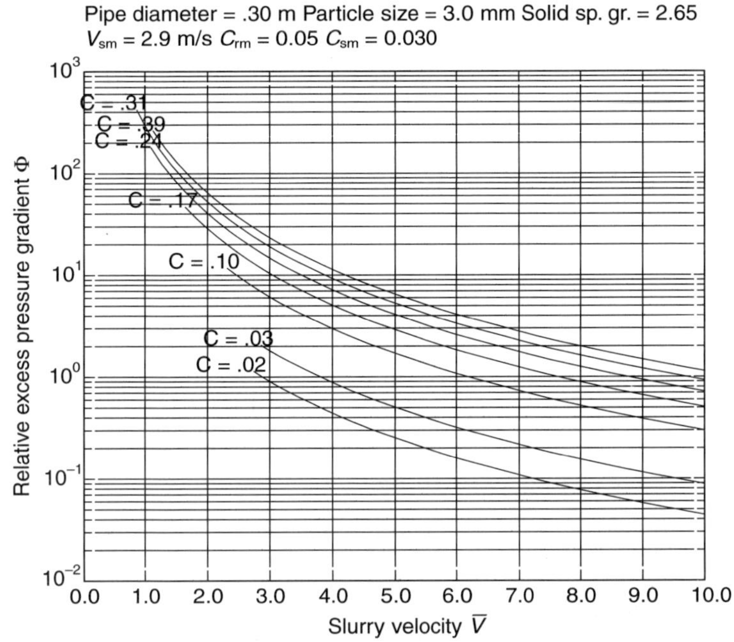

Figure 4.18 Relative excess pressure gradient for a sand slurry in heterogeneous suspension in a 30 cm pipe. $d_{50} = 3\,\text{mm}$, $d_{85} = 6\,\text{mm}$

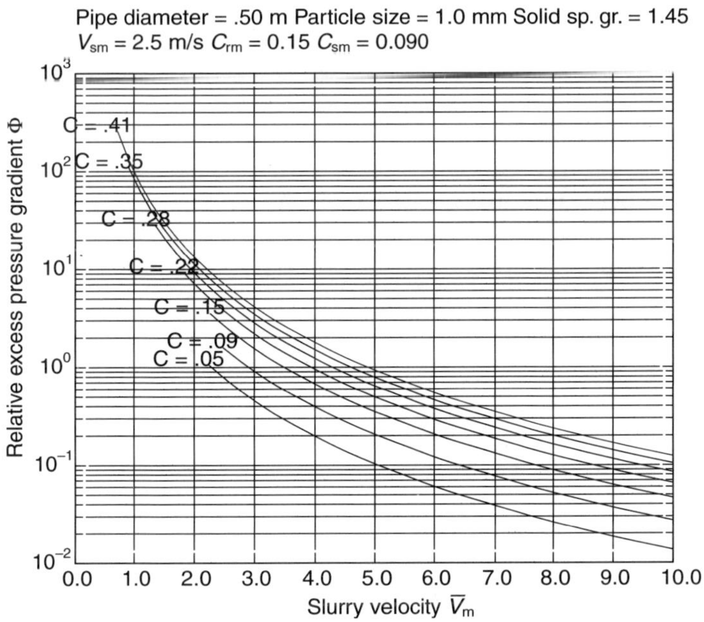

Figure 4.19 Relative excess pressure gradient for a coal slurry in heterogeneous suspension in a 50 cm pipe. $d_{50} = 1\,\text{mm}$, $d_{85} = 2\,\text{mm}$

Illustrative Example 4.7 Heterogeneous slurry flow

Calculate the pressure gradient due to friction when a slurry of sand in water having $d_{50} = 0.63\,\text{mm}$ and $d_{85} = 0.74\,\text{mm}$ is transported through a 20.3-cm horizontal pipe at a solid volume fraction of 0.138. The density of the sand is $2650\,\text{kg/m}^3$, and the slurry flows at $3\,\text{m/s}$. Assume that the coefficient of friction between the settled solids and the wall is 0.44.

Use the FLUIDS toolbox to get the terminal settling velocities for particles of size 0.63 mm and 0.74 mm, respectively.

$$v_T = 0.104\,\text{m/s} \quad \text{for} \quad d_p = 0.63\,\text{mm}$$

$$v_T = 0.123\,\text{m/s} \quad \text{for} \quad d_p = 0.74\,\text{mm}$$

$$w_{50} = 0.9 \times 0.104 + 2.7\left[\frac{(2650 - 1000) \times 9.81 \times 0.001}{1000^2}\right]^{1/3}$$

$$= 0.094 + 0.068 = 0.162$$

$$w_{85} = 0.9 \times 0.123 + 0.068 = 0.179$$

Use the FLUIDS toolbox to get the friction factor for the carrier fluid.

$$Re = \frac{0.203 \times 3 \times 1000}{0.001} = 6.09 \times 10^5$$

$$f_w = 0.00307$$

$$\bar{V}_{50} = 0.162\left(\frac{2}{0.00307}\right)^{1/2} \cosh\left(60\frac{0.00063}{0.203}\right)$$

$$= 4.207\,\text{m/s}$$

$$\sigma_s = 0.046$$

$$M = \min\,[0.25 - 13 \times 0.046^2)^{-1/2}, 1.7] = 1.7$$

$$\Phi = \frac{0.138(2.65 - 1) \times 0.44 \times 9.81 \times 0.203}{4 \times 0.00307 \times 3^2}\left(\frac{4.207}{3}\right)^{1.7}$$

$$= 3.208$$

$$-\frac{\Delta P_{fw}}{L} = \frac{2 \times 0.00307 \times 1000 \times 3^2}{0.203}$$

$$= 272\,\text{Pa/m}$$

$$-\frac{\Delta P_{fsl}}{L} = 272(1 + 3.208)$$

$$= 1.14\,\text{kPa/m}$$

The FLUIDS toolbox can be used to obviate the need for all the calculation that is associated with these empirical expressions. The data input screen is shown

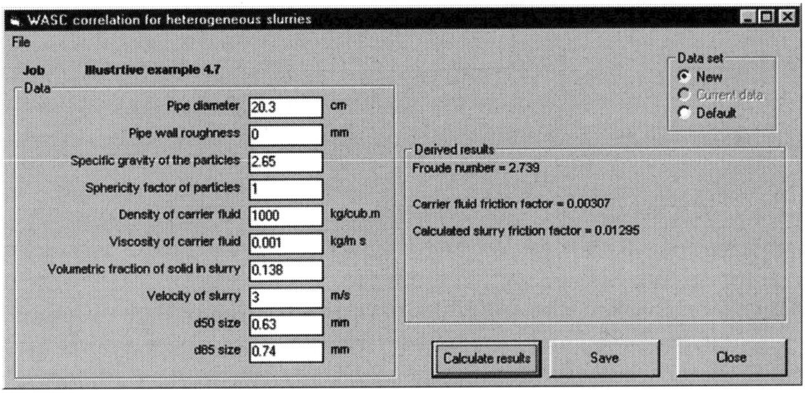

Figure 4.20 Data input screen for calculating the friction factor for a slurry in heterogeneous suspension

in Figure 4.20. The method is referred to as the WASC correlation because it was developed by Wilson *et al.* (1997).

4.5 Flow of settling slurries in vertical pipelines

When the flow is vertical, the solids settle in a direction parallel to the average direction of motion of the slurry. As a result the collisions between particles and wall are very much less frequent than in horizontal pipelines. Thus the additional momentum transfer paths shown in Figure 4.1 can be neglected and the entire energy dissipation is due to the frictional drag of the carrier fluid against the pipe wall. However, the actual carrier fluid velocity must be used when calculating the energy dissipated by viscous shear and this velocity is not equal to the average slurry velocity as is shown below.

The energy dissipated per unit mass of slurry in a segment of vertical pipe is calculated as follows. Consider a segment of the pipe that contains *m* kg of slurry. The volume of this segment is $A_c L$ where A_c is the cross-sectional area of the pipe. The frictional dissipation of energy per unit mass of slurry is calculated as the product of the force acting on the slurry multiplied by the distance moved.

$$F = \frac{-\Delta p_{\mathrm{f,sl}} A_c L}{m} \quad \text{J/kg slurry} \tag{4.51}$$

$$\rho_{\mathrm{sl}} = \frac{m}{A_c L}$$

Since

$$F = \frac{-\Delta p_{\mathrm{f,sl}}}{\rho_{\mathrm{sl}}} \quad \text{J/kg slurry} \tag{4.52}$$

The pressure drop is calculated from the viscous shear of the water against the pipe wall.

$$-\Delta P_{f,sl} = 2\rho_w \bar{V}_w^2 f_w \frac{L}{D} \tag{4.53}$$

$$F = 2\bar{V}_w^2 f_w \frac{L}{D}\frac{\rho_w}{\rho_{sl}} \tag{4.54}$$

f_w is obtained from the friction factor chart using the actual velocity of the water $\bar{V}_w$ to evaluate the Reynolds number.

Here ρ_{sl} is the actual density in the vertical leg. This will be different to the average density of the slurry in any horizontal sections of the pipeline and in the slurry that discharges from the pipeline system because solid particles will accumulate in vertical pipes in which the flow is upward and will be swept out of vertical pipes in which the flow is downward.

The water flows through the interstices of the slurry and the frictional drag on the wall is due to the liquid alone.

4.5.1 Water velocity and slurry composition in a vertical pipe

Because the solid settles relative to the water at the local settling velocity, the velocity of the water is greater than that of the solid and greater than the mean velocity of the slurry in segments of the pipeline where the flow is upward. This greater velocity of the water must be used in assessing the pressure drop of the liquid. In the following, it is assumed that the local settling velocity is equal to the terminal settling velocity. This assumption may severely restrict the validity of the method when flow is turbulent but the assumption is ameliorated somewhat by the use of only average velocities rather than the local turbulent eddy velocity.

The volume fraction of solids in the vertical section, q, is different from that fed to or removed from the system because of the slip velocity. The actual velocity of the water and the actual volume fraction of solid must be calculated. This can be done using a calculation based on the material balance for the slurry in the vertical pipe.

Let

ν_T = settling velocity of the solids relative to the liquid.
$\bar{V}_w$ = water velocity
$\bar{V}$ = mean velocity of the slurry.

Consider any cross-section of a vertical segment:
Water volumetric flowrate

$$\bar{V}_w(1-q) \times A_c \quad \text{m}^3/\text{s}$$

Solid volumetric flowrate

$$(\bar{V}_w - \nu_T)q \times A_c \quad \text{m}^3/\text{s}$$

Thus

$$\bar{V} = \bar{V}_W(1-q) + (\bar{V}_W - \nu_T)q \tag{4.55}$$

$$\bar{V} = \bar{V}_W - \nu_T q \tag{4.56}$$

The volume fraction of solids leaving the system is given by

$$C = \frac{\text{solid flowrate}}{\text{total flowrate}} = \frac{(\bar{V}_W - \nu_T)q}{\bar{V}_W(1-q) + (\bar{V}_W - \nu_T)q} \tag{4.57}$$

$$= \frac{(\bar{V}_W - \nu_T)q}{\bar{V}_W - \nu_T q} \tag{4.58}$$

Any two of the four variables can be calculated from equations 4.56 and 4.58 if the other three are known. Typically $\bar{V}$, C and ν_T are known from which V_W and q can be calculated.

Once V_W is known the total pressure gradient can be calculated using equation 4.53.

The effective density of the slurry in the vertical pipe segment is

$$\bar{\rho}_{\text{slurry}} = q\,\rho_s + (1-q)\,\rho_l \tag{4.59}$$

$$= \rho_l + q(\rho_s - \rho_l) \tag{4.60}$$

The following cases are important in practice:

1. C and $\bar{V}$ are known: This occurs when a known quantity of slurry must be pumped through a pipe of known diameter. In this situation q and $\bar{V}_W$ are the unknowns that must be calculated.

Equation 4.58 is written

$$\bar{V}_W(q - C) = q\,\nu_T(1 - C) \tag{4.61}$$

Substituting for $\bar{V}_W$ from equation 4.56

$$(\bar{V} + q\nu_T)(q - C) = q\nu_T(1 - C) \tag{4.62}$$

which is quadratic in q and the solution is

$$2q = \left(1 - \frac{\bar{V}}{\nu_T}\right) \pm \sqrt{\left(1 - \frac{\bar{V}}{\nu_T}\right)^2 + 4C\frac{\bar{V}}{\nu_T}} \tag{4.63}$$

and

$$\bar{V}_W = \bar{V} + \nu_T q \tag{4.64}$$

Note that $\bar{V}$ and V_W are negative when the flow is downward and the negative sign is used in front of the root in equation 4.63.

2. $\bar{V}_W$ is known: This applies when a minimum water velocity is required to lift particles of a given size and type such as the largest and heaviest

particles that are expected to occur in the slurry. In this situation q and $\bar{V}$ are the unknown quantities that must be calculated

From equation 4.58

$$q = \frac{C\,\bar{V}_{\mathrm{w}}}{\bar{V}_{\mathrm{w}} - \nu_{\mathrm{T}} + C\,\nu_{\mathrm{T}}} \tag{4.65}$$

and

$$\bar{V} = \bar{V}_{\mathrm{w}} - \nu_{\mathrm{T}}\, q \tag{4.66}$$

The required pipe diameter and the frictional dissipation of energy can be calculated using equations 4.14 and 4.54.

4.5.2 The effect of particle size- and density-distribution in vertical pipes

The particle population will exhibit a distribution of particle sizes and particle density. The principle to be adopted under these conditions is that each particle type has its own terminal settling velocity ν_{Ti} and will find its own value of q_i and the q_i's are additive. Let particle type i be present in the vertical pipe at volume concentration q_i. The total volumetric concentration of solid q is given by

$$q = \sum_i q_i \tag{4.67}$$

Volumetric flowrate of the water $= \bar{V}_{\mathrm{w}}\,(1 - q)\,A_{\mathrm{c}}$ and the volumetric flowrate of the solid is given by

$$\sum_i (\bar{V}_{\mathrm{w}} - \nu_{\mathrm{Ti}})\, q_i\, A_{\mathrm{c}} \tag{4.68}$$

The total volumetric flowrate is influenced by the settling velocities of all the particles and using equation 4.55

$$\begin{aligned} \bar{V} &= \bar{V}_{\mathrm{w}}(1 - \sum_i q_i) + \sum_i (\bar{V}_{\mathrm{w}} - \nu_{\mathrm{T}i})q_i \\ &= \bar{V}_{\mathrm{w}} - \sum_i \nu_{\mathrm{T}i} q_i \end{aligned} \tag{4.69}$$

The volumetric fraction of solid particles of type i in the slurry is given by

$$\begin{aligned} C_i &= \frac{\text{Solid flowrate (type } i)}{\text{Total flowrate}} \\ &= \frac{(\bar{V}_{\mathrm{w}} - \nu_{\mathrm{T}i})q_i}{\bar{V}} \end{aligned} \tag{4.70}$$

The two important types of problem are solved as follows.

1. $\bar{V}$ and C_i are known: this situation arises when the nature of the slurry is known and the quantity that must be transported is fixed.

From equation 4.69

$$\bar{V}_{\mathrm{W}} = \bar{V} + \sum_i \nu_{\mathrm{T}i}\, q_i \tag{4.71}$$

and from equation 4.70

$$\begin{aligned} C_i \bar{V} &= \bar{V}_{\mathrm{W}} q_i - \nu_{\mathrm{T}i} q_i \\ &= \bar{V} q_i + q_i \sum_j \nu_{\mathrm{T}j} q_j - \nu_{\mathrm{T}i} q_i \end{aligned} \tag{4.72}$$

Equation 4.72 is true for each particle type and equations 4.71 and 4.72 give as many equations as there are unknowns. These are solved most conveniently by iteration and the following procedure has been found to be effective and rapidly convergent. Add each of the equations 4.72

$$\begin{aligned} \sum_i C_i \bar{V} &= \bar{V} \sum q_i + \sum_i q_i \sum_j \nu_{\mathrm{T}j} q_j - \sum \nu_{\mathrm{T}i} q_i \\ &= \bar{V} q + (q-1) \sum_i \nu_{\mathrm{T}i} q_i \end{aligned} \tag{4.73}$$

From which

$$\sum_i \nu_{\mathrm{T}i} q_i = \frac{\bar{V}(C-q)}{q-1} \tag{4.74}$$

The volumetric fraction for species i is obtained from equation 4.70 as

$$\begin{aligned} q_i &= \frac{C_i \bar{V}}{\bar{V} + \sum_i \nu_{\mathrm{T}i} q_i - \nu_{\mathrm{T}i}} \\ &= \frac{C_i \bar{V}}{\bar{V} + \frac{\bar{V}(C-q)}{q-1} - \nu_{\mathrm{T}i}} \\ &= \frac{C_i}{1 + \frac{C-q}{q-1} - \frac{\nu_{\mathrm{T}i}}{\bar{V}}} \end{aligned} \tag{4.75}$$

A convenient iterative solution starts with an estimate of q from which the q_i's are calculated using equation 4.75. The assumed value of q is checked using $q = \sum q_i$ which also provides a new refined starting value for the next iteration.

At convergence the velocity of the water can be recovered from equations 4.71 and 4.74

$$\bar{V}_{\mathrm{W}} = \bar{V}\left(1 + \frac{C-q}{q-1}\right) = \bar{V}\frac{1-C}{1-q} \tag{4.76}$$

and the frictional dissipation of energy can be calculated from $\bar{V}_{\mathrm{W}}$ and the friction factor for the carrier fluid.

2. $\bar{V}_{\mathrm{w}}$ and C are known.

When $\bar{V}_{\mathrm{w}}$ and C are known, a similar iterative solution is possible.
From equation 4.72

$$C_i \bar{V} = \bar{V}_{\mathrm{w}} q_i - \nu_{\mathrm{T}i} q_i \tag{4.77}$$

and substituting for $\bar{V}$ from equation 4.69

$$C_i \left(\bar{V}_{\mathrm{w}} - \sum_j \nu_{\mathrm{T}j} q_j \right) = \bar{V}_{\mathrm{w}} q_i - \nu_{\mathrm{T}i} q_i \tag{4.78}$$

By a process similar to that used to develop equation 4.75 it is easy to show that

$$q_i = \frac{C_i - \dfrac{C - q}{C - 1}}{1 - \dfrac{\nu_{\mathrm{T}i}}{\bar{V}_{\mathrm{w}}}} \tag{4.79}$$

This equation can be used in place of equation 4.75 to generate the iterative process.

At convergence the average slurry velocity can be calculated from equation 4.77 and then the required pipe diameter.

Use the FLUIDS toolbox to implement these solution methods.

4.6 Practice problems

1. Calculate the flowrate of slurry through 200 m of smooth horizontal pipe, having internal diameter 13 cm, under a pressure drop of 5.4 bar.
   The friction factor for water can be calculated from $f_{\mathrm{w}} = 0.079Re^{-0.25}$
   The Durand–Condolios–Worster correlation can be used for the slurry.
   Data:

   Particle size 100 μm.

   Particle density 2670 kg/m^3

   Density of water 1000 kg/m^3

   Viscosity of water 0.001 kg/ms

   Slurry is 30 per cent by volume

   1 bar = 10^5 Pa

   Use the FLUIDS toolbox to calculate the flowrate using the Turian–Yuan correlations.
2. The following expressions have been found to apply for calculating the pressure drop when a wood-chip slurry is pumped through a horizontal pipeline. Calculate the power required per meter of pipe length to pump

the slurry at a rate of 0.003 m^3/s. The diameter of the pipe should be chosen to minimize the pressure drop.

$$\frac{\Delta P_{f,sl} - \Delta P_w}{\Delta P_w} = 2.51C\left(\frac{4gD}{\bar{V}^2}\right)^{1.42}$$

$$f_w = 0.079Re^{-0.25}$$

The slurry contains 20 per cent wood by volume and has density 980 kg/m^3.

3. The New Zealand Steel iron sand slurry pipeline has the following design specifications:

   Operating velocity = 3.9 m/s

   Concentration = 48.5 wt per cent solids

   Solid specific gravity = 4.76

   Mean particle size = 120 μm

   Pipeline inside diameter = 187.4 mm

   The pipeline is constructed in two sections of 9.2 km and 8.8 km respectively.
   Calculate the design tonnage for solids delivered by the pipeline.
   Calculate the pressure gradient due to friction and the power consumption required to overcome friction when the pipeline operates at the design capacity.
   Use the following values for the properties of water
   Density of water = 1000 kg/m^3
   Viscosity of water = 0.001 kg/ms

3. 4410 t/h of sand are to transported as a slurry in a 25.6 inch ID pipeline at a concentration of 20 by volume.
   Calculate the required slurry velocity and the pressure gradient due to friction.
   Data:

   Density of the sand = 2700 kg/m^3.

   Density of water = 996 kg/m^3.

   Viscosity of water = 0.798 cP.

   d_{50} for the sand = 0.70 mm.

   d_{85} for the sand = 1.00 mm.

4. Your company wishes to design a hydraulic transport system for waste metal stampings in the form of small metal platelets. Two shapes are made in the ratio 33 to 67 by mass. These are identified as types A and B respectively. Terminal settling velocities for the two shapes were measured and are 199 mm/s and 227 mm/s respectively. A 47 per cent by volume slurry of the stampings is pumped upward through a vertical

pipeline at an average velocity of 1.7 m/s. The specific gravity of the platelets is 7.52 and the sphericity factor for both types is 0.53. Calculate the actual concentration of platelets in the vertical pipeline and calculate the pressure gradient due to friction.

Pipe diameter is 2.5 cms

Density of water = 1000 kg/m^3

Viscosity of water = 0.001 kg/ms

Choose a suitable pump for this application and specify the pump motor power required if the stampings must be lifted through 11 m. What quantity of stampings can be transported?

5(a). Calculate the rate of energy dissipation due to friction when 800 m^3/hr of water is pumped through 100 m of smooth 20.3 cm ID smooth pipe. Data:

Density of water = 1000 kg/m^3.

Viscosity of water = 0.001 kg/ms.

5(b). A pump having generalized pump characteristic constants $A = 5.8$, $B = -11.0$, $C = 700$ and impeller diameter 55 cm is used to pump 800 m^3/hr of water. What is the theoretical power required by the pump if it runs at 1200 rpm?

5(c). At what speed must the pump run to deliver 800 m^3/hr into a piping system that is equivalent to 100 m of smooth 20.3 cm internal diameter pipe with a static head of 23 m?

5(d). What power does the pump draw when running under the conditions specified in section 5(c) above?

6. Repeat questions 5(a)–5(d) but the fluid is now a settling slurry containing 43 per cent by mass of spherical silica particles of size 0.5 mm. In part (a) the pipe may be considered to be horizontal over its whole length. In part (c) the pipe consists of 77 m of horizontal and 23 m of vertical pipe. The density of silica is 2700 kg/m^3.

7. All possible regime boundaries calculated for a slurry in a horizontal pipeline are shown in Figure 4.9. Mark all lines in the figure that represent real regime boundaries.

8(a). The two vertical sections in laboratory pumping loop are 294 cm long. The fluid under test flows upward in one leg and downward in the other. Pressure gauges are installed at the top and bottom of each leg. Calculate the difference in pressure registered by the pressure gauges across each leg when water flows in the loop at 2.19 m/s. The internal diameter of the pipe is 2.60 cm.

8(b). Repeat the calculation of section 8(a) when a limestone slurry flows through the loop. The slurry contains 42 per cent by mass limestone which has a density of 2730 kg/m^3. The particle size distribution is given in Table 4.1.

Table 4.1 Particle size distribution of limestone

Mesh size (mm)	*passing (%)*
7.14	99.0
3.57	86.5
1.78	58.2
0.631	22.2
0.112	3.1

4.7 Symbols used in this chapter

A_c	Cross-sectional area of pipe m^2.
C	Volume fraction of solids in slurry. This is the concentration that is discharged from the pipeline. Concentrations in individual segments of the pipe may be different. Volume fraction.
C_D^*	Drag coefficient at terminal settling velocity.
C_r	Solid concentration relative to C_{vb}.
C_{rm}	Critical deposit concentration relative to C_{vb}.
C_{sm}	Critical deposit concentration.
C_{vb}	Volume fraction of solids for loosely packed bed.
D	Pipe diameter m.
F	Energy dissipated by friction J/kg.
F_{fr}	Frictional force between settled bed and pipe wall *N*.
F_N	Normal force exerted by settled bed on pipe wall *N*.
Fr	Froude number.
f_{sl}	Friction factor for the slurry.
f_w	Friction factor for the carrier fluid.
g	Acceleration due to gravity m/s^2.
i	Slurry pressure gradient (hydraulic gradient) m water/m pipe.
i_{pg}	Slurry pressure gradient when flowing as a plug at concentration C_{vb}.
j	Slurry pressure gradient m slurry/m of pipe.
L	Pipe length m.
q	Concentration of slurry in a vertical segment of the pipe. Volume fraction.
Q	Volumetric flowrate of slurry m^3/s.
R_{ab}	Transition number between regimes *a* and *b*.
s	Specific gravity of solid.
$\bar{V}$	Average velocity of slurry in pipe m/s.
V_s	Limiting velocity for stationary deposition in a horizontal pipe m/s.
V_{sm}	Maximum value of V_s m/s.
$\bar{V}_w$	Average water velocity in vertical pipe segment m/s
v_T	Terminal settling velocity m/s.
$\Delta P_{f,w}$	Pressure drop due to friction for carrier fluid alone Pa.
ζ	Excess pressure gradient relative to i_{pg}.

μ_s Coefficient of friction between settled bed and pipe wall.
μ_w Viscosity of carrier fluid Pa s.
ρ_s Density of solid kg/m^3.
ρ_{sl} Density of slurry kg/m^3.
ρ_w Density of carrier fluid kg/m^3.
Φ Fractional increase in frictional pressure drop due to presence of particles.
Ω Constant in Durand–Condolios–Worster correlation.

Bibliography

The use of the concept of excess pressure gradient as a useful correlating variable originated with Durand in the early 1950s. The inclusion of the effect of particle density is attributed to Worster (Smith 1955, Bain and Bonnington, 1970). The uncertainty in the results that are obtained from the original Durand–Condolios correlation has been frequently discussed in the literature, for example in Govier and Aziz (1972, Chapter 11).

The analysis using separate correlations for the four flow regimes is due to Turian and Yuan (1977).

The stratified flow model is based on the work of Wilson *et al.* (1997). Their comprehensive text covers many aspects of slurry transport and is recommended to the student particularly for the many interesting and educational case studies.

The treatment of flow of settling slurries in vertical pipes is based on Bain and Bonnington (1970).

References

Bain, A.G. and Bonnington, S.T. (1970). *The Hydraulic Transport of Solids by Pipeline.* Pergamon Press.

Govier, G.W. and Aziz, K. (1972). *The Flow of Complex Mixtures in Pipes.* Van Nostrand Reinhold Co.

Smith, R.A. (1955). Experiments on the flow of sand-water slurries in horizontal pipes. *Trans. Instn. Chem. Engnrs* **33**, 85–92.

Turian, R.M. and Yuan, T.-F. (1977). Flow of Slurries in Pipelines. *AIChE Journal* **23**, 232–243.

Wilson, K.C., Addie, G.R., Sellgren, A. and Clift, R. (1997). *Slurry Transportation using Centrifugal Pumps* 2nd edition. Blackie Academic and Professional.

5
Non-Newtonian slurries

5.1 Rheological properties of fluids

When the concentration of solid in a slurry is greater than about 50 per cent by volume, the slurry changes its flow characteristics and the Newtonian behavior of the carrier fluid no longer dominates the flow as it does in the case of the settling slurries that are discussed in Chapter 4. The internal momentum transfer processes must reflect the role that the densely packed solid particles play. If the particles are in the colloidal or micron size range, the slurries deviate from Newtonian behavior and they require more complex rheological models to describe their flow behavior. The relationship between the local rate of deformation of the fluid and the shearing stress that is imposed differs from that which characterizes Newtonian fluids. This relationship is used to classify the flow behavior of various non-Newtonian fluids.

A Newtonian fluid is characterized by a linear relationship between the local shearing stress and the rate of strain within the moving fluid.

$$\tau = -\mu \frac{\mathrm{d}u}{\mathrm{d}r} \tag{5.1}$$

The proportionality constant is the viscosity of the fluid. $\mathrm{d}u/\mathrm{d}r$ is the local spatial gradient of the velocity in the fluid. Newtonian fluids have constant viscosity at all stresses and shear rates.

Non-Newtonian fluids exhibit various types of non-linearity. Four distinct types of non-Newtonian behavior are illustrated in Figure 5.1 which shows how the rate of strain $\mathrm{d}u/\mathrm{d}r$ varies with the shear stress that is applied to the fluid. The relationship between shear stress and rate of strain for fluids can be measured in the laboratory and careful measurements can establish the rheological characteristic of any particular fluid. Dense slurries made from fine particles often behave as Bingham plastics at least approximately and this is a useful model for these fluids. The Bingham plastic is an idealized model that can be used in theoretical calculations but no real fluids ever behave exactly as a Bingham plastic. Figure 5.2 shows experimentally determined rheological behavior for two mineral slurries and it may reasonably be assumed that the Bingham plastic model is an adequate model to describe the flow behavior of these slurries.

5.1.1 Bingham plastic fluids

Bingham plastics exhibit a linear relationship between shear stress and rate of strain but, unlike Newtonian fluids, this relationship does not exhibit a zero intercept at zero strain rate. The intercept at zero strain rate is called the yield

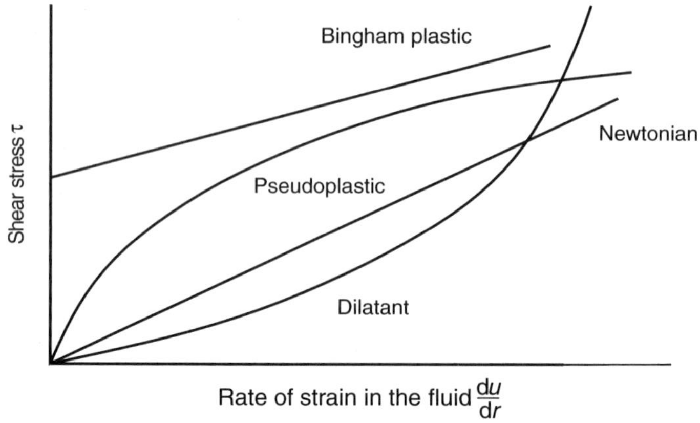

Figure 5.1 Typical stress-rate-of-strain relationship for non-Newtonian fluids

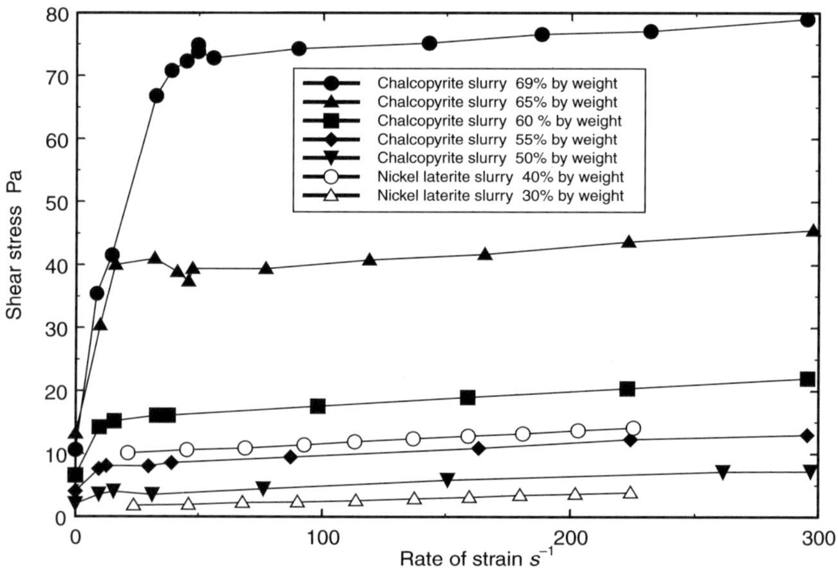

Figure 5.2 Viscometer data for two mineral slurries showing approximate Bingham plastic behavior. Data from Huyhn *et al.* (2000) (closed symbols) and Bhattacharya *et al.* (1998) (open symbols)

stress because it is the minimum shear stress that must be applied to the fluid before it will deform at all.

The rheological model for the Bingham plastic is

$$\begin{aligned} -\mu_B \frac{du}{dr} &= \tau - \tau_Y \quad \text{when } \tau \geq \tau_Y \\ \frac{du}{dr} &= 0 \quad \text{when } \tau < \tau_Y \end{aligned} \tag{5.2}$$

μ_B is called the coefficient of rigidity or coefficient of plastic viscosity. τ_Y is the yield stress. This model is based on the idea that the fluid will not deform, and therefore $(du/dr) = 0$, unless the shear stress acting on the fluid exceeds a definite critical yield stress τ_Y. The parameters μ_B and τ_Y that describe the behavior of these slurries vary with the solids concentration, the particle size, particle shape, and specific surface area and the chemical conditions of the particle surfaces. The concentration of any potential-determining ions often has a significant effect on these parameters. For example H^+ and OH^- ions are potential determining for many minerals and significant variations in the values of μ_B and τ_Y with pH have often been reported.

This type of behavior is explained physically by the fluid having a 3-D structure associated with the densely packed solids that resists deformation unless the stress is high enough to break down this structure. Once the structure has been loosened the fluid exhibits Newtonian behavior with a linear relationship between the shear stress and the rate of strain. In the absence of any hysteresis effects, the structure is reconstituted as soon as the stress is decreased below τ_Y. It is not likely that any real fluid will exhibit Bingham plastic behavior exactly but the model is a useful approximation for many real dense slurries. Some typical data measured in laboratory viscometers is shown in Figure 5.2.

5.1.2 Pseudo plastic fluids

These fluids are characterized by two distinct effective viscosities at the extremes of low stress and high stress with a smooth transition in between. This implies that any structural constraints in the fluid are smoothly overcome as the stress increases and as particles and molecules are aligned more or less in the direction of flow. The effective viscosity decreases as the shear stress increases.

The effective viscosity is defined to be the ratio of the shear stress to the rate of strain

$$\mu_{\text{eff}} = \frac{\tau}{\left|\dfrac{du}{dr}\right|} \tag{5.3}$$

Three simple models for pseudo plastic behavior that use the concept of effective viscosity have been used to describe the rheological behavior of some non-Newtonian fluids:

1. Meter Model:

$$\mu_{\text{eff}} = \mu_\infty + \frac{\mu_0 - \mu_\infty}{1 + \left(\dfrac{\tau}{\tau_m}\right)^{\alpha-1}} \tag{5.4}$$

The constants in this model have the following significance

μ_0 = viscosity at low shear rate,

μ_∞ = viscosity at high shear rate,

τ_m = the shear stress at which the effective viscosity lies midway between μ_0 and μ_∞,
α = a parameter that describes how quickly the effective viscosity varies from its lower shear-rate limit to its high shear-rate value.

2. Seely Model:

$$\mu_{\text{eff}} = \mu_\infty + (\mu_0 - \mu_\infty)e^{-\gamma\tau} \tag{5.5}$$

This model postulates an exponential decrease in the effective viscosity as the shear stress increases.

3. Ostwald-de Waele or Power Law Model:

$$\mu_{\text{eff}} = m^\alpha |\tau|^{1-\alpha} \tag{5.6}$$

or

$$\tau = K\left(-\frac{du}{dr}\right)^n \tag{5.7}$$

Since du/dr is negative in conduits of circular cross-section, the term inside the brackets is positive which is essential since n is usually not an even integer. The parameter K in equation 5.7 is called the fluid consistency coefficient and n the flow behavior index.

These models are all based on the idea that the effective viscosity varies as the local rate of strain in the fluid increases. If the effective viscosity decreases as the rate of strain increases ($\mu_0 < \mu_\infty$ or $n < 1$) the fluid is shear thinning and behaves as a pseudo plastic. If the effective viscosity increases with increasing strain rate ($\mu_0 > \mu_\infty$ or $n > 1$) the fluid is shear thickening or dilatant.

Some fluids exhibit rheological behavior that appears to include characteristics of Newtonian, yield stress and power law behavior to varying degrees as the local rate of strain varies. Rather more complex models are required to describe the behavior of such fluids and models which are constructed as combinations of the simpler models have been found to be useful.

5.1.3 Pseudo plastic fluids with yield stress

Fluids which exhibit a yield stress but also exhibit a non-linear relationship between shear stress and local rate of strain can be modeled using the equation

$$\begin{aligned} K_H\left(-\frac{du}{dr}\right)^n &= \tau - \tau_H \quad \text{when} \quad \tau \geq \tau_H \\ &= 0 \quad \text{when} \quad \tau < \tau_H \end{aligned} \tag{5.8}$$

This is commonly referred to as the Herschel–Bulkley model.

A non-linear version of this is the Casson model

$$K_c\left(-\frac{du}{dr}\right)^{0.5} = \tau^{0.5} - \tau_C^{0.5} \quad \text{when} \quad \tau \geq \tau_C$$
$$= 0 \qquad \text{when} \quad \tau < \tau_C \tag{5.9}$$

τ_C is the Casson yield stress.

5.1.4 Shear-thinning fluids with Newtonian limit

Some shear-thinning fluids exhibit Newtonian behavior at high strain rates. The rheological properties of these fluids can be modeled using

$$\tau = \mu_\infty\left(-\frac{du}{dr}\right) + K_S\left(-\frac{du}{dr}\right)^n \tag{5.10}$$

As the strain rate becomes large

$$\tau = \mu_\infty\left(-\frac{du}{dr}\right) \tag{5.11}$$

provided that $n < 1$ which shows that the fluid exhibits Newtonian behavior at very high strain rates. This is commonly referred to as the Sisko model.

Typical data measured using slurries of fine TiO_2 are shown in Figure 5.3. The shear stress calculated using equation 5.10 is shown as a series of solid

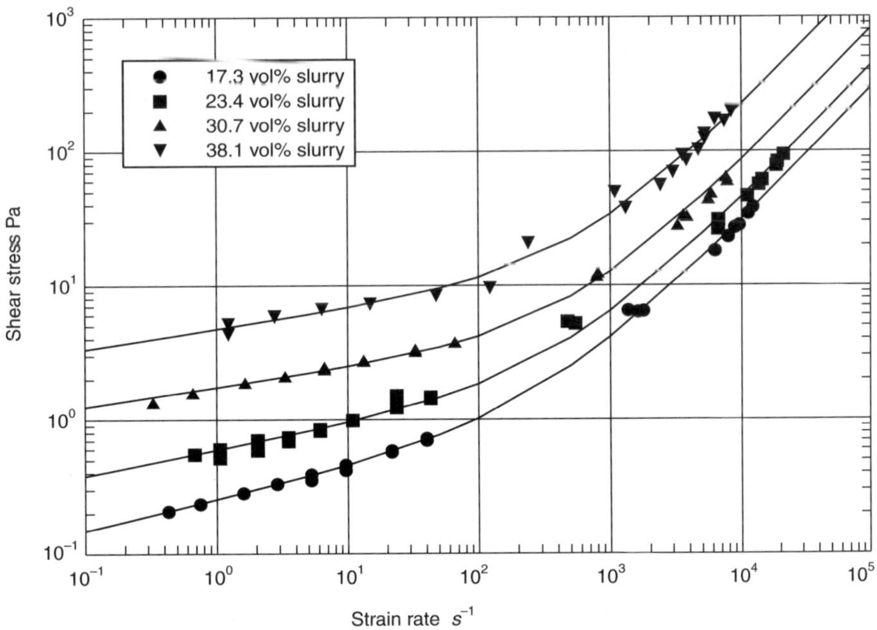

Figure 5.3 Measured shear stress–rate of strain relationships for TiO_2 slurries. Data from Ma (1987)

Table 5.1 Sisko model parameters for TiO_2 slurries

Slurry (volume %)	*High-shear viscosity* μ_∞ Pa s	*Consistency coefficient* K_S Pa s^n	*Flow index n*
17.3	0.00289	0.252	0.231
23.4	0.00429	0.591	0.191
30.7	0.00800	1.736	0.145
38.1	0.02070	4.729	0.150

lines in Figure 5.3. The values of the three rheological parameters in the Sisko model for the four slurries are given in Table 5.1.

The parameters in all of the models described in the previous sections should be regarded as empirical curve fitting constants. They do not have any real physical significance in terms of the fundamental molecular and physical properties of the fluids they are meant to describe. Consequently, they must be obtained from careful laboratory experiments in which the shear stress and rate of strain are measured. It is difficult to establish, with reasonable confidence, independent estimates of the values of as many as three parameters in any particular model from experimental data and for this reason two-parameter models are generally preferred over the more elaborate three- or four-parameter models. The measurement of rheological properties falls outside of the scope of this book and the reader is referred to the bibliography at the end of this chapter for sources of information on this topic.

5.2 Newtonian and non-Newtonian fluids in pipes with circular cross-section

Because of the inherent structure that is exhibited by non-Newtonian slurries, laminar motion of these fluids is encountered far more commonly than with Newtonian fluids. It is therefore profitable to examine some simple cases of laminar flow which can be analyzed completely by theoretical methods.

Consider a viscous fluid (Newtonian or non-Newtonian) flowing through a round tube as illustrated in Figure 5.4.

Figure 5.4 A force balance over an imaginary cylinder can be used to calculate the velocity profile and the average velocity of the fluid as a function of the pressure gradient due to friction

A force balance on an imaginary cylindrical core of radius r in the center of the tube gives

$$-\Delta P_f \pi r^2 = \tau 2\pi r \Delta x \tag{5.12}$$

$$\tau = -\frac{1}{2}\frac{\Delta P_f}{\Delta x} r \tag{5.13}$$

which shows that the shear stress in the fluid varies linearly with the radius.

At the wall $\tau = \tau_w$ and $r = D/2$
so that

$$\tau_w = -\frac{1}{2}\frac{\Delta P_f}{\Delta x}\frac{D}{2} \tag{5.14}$$

and

$$\frac{\tau}{\tau_w} = \frac{2r}{D} \tag{5.15}$$

This equation shows that the shear stress varies linearly with radius within any fluid that flows inside a circular pipe. The equation has wide applicability and is not limited to Newtonian or even to viscous fluids. It applies to turbulent flow as well as laminar flow conditions. This equation is useful for developing a relationship between the average velocity and the frictional pressure gradient for a wide variety of fluids.

The average velocity across the pipe diameter can be calculated by integration of the velocity profile.

$$\begin{aligned}\bar{V} &= \frac{8}{D^2}\int_0^{D/2} ur\,\mathrm{d}r \\ &= \frac{4}{D^2}\int_0^{D/2} u\frac{\mathrm{d}r^2}{\mathrm{d}r}\mathrm{d}r\end{aligned} \tag{5.16}$$

Here u is the velocity of the fluid in the axial direction at radius r.

By noting that $u = 0$ at $r = D/2$, equation 5.16 can be integrated by parts to give

$$\bar{V} = \frac{4}{D^2}\int_0^{D/2} r^2\left(-\frac{\mathrm{d}u}{\mathrm{d}r}\right)\mathrm{d}r \tag{5.17}$$

This form of the equation is useful because the various rheological models of the fluid will relate the velocity gradient $\mathrm{d}u/\mathrm{d}r$ to the shear stress when the flow is laminar. The shear stress in turn varies linearly with radius according to equation 5.15.

Using equation 5.15 the variable of integration in equation 5.17 can be changed to τ.

$$\bar{V} = \frac{D}{2\tau_w^3}\int_0^{\tau_w} \tau^2\left(-\frac{\mathrm{d}u}{\mathrm{d}r}\right)\mathrm{d}\tau \tag{5.18}$$

or more generally

$$\bar{V} = \int_0^1 \xi^2 \left(-\frac{\mathrm{d}u}{\mathrm{d}\xi} \right) \mathrm{d}\xi \tag{5.19}$$

with $\xi = 2r/D = \tau/\tau_\mathrm{w}$ representing a dimensionless radius or dimensionless stress as convenient.

The shear stress at the wall can be related to the frictional pressure drop and to the friction factor using equation 5.14

$$\begin{aligned} \tau_\mathrm{w} &= \frac{D}{4}\frac{-\Delta P_\mathrm{f}}{L} \\ &= \frac{1}{2}\rho_\mathrm{sl}\bar{V}^2 f_\mathrm{sl} \end{aligned} \tag{5.20}$$

This is equivalent to

$$PGDTF = \frac{-\Delta P_\mathrm{f}}{L} = \frac{2\rho_\mathrm{sl}\bar{V}^2 f_\mathrm{sl}}{D} \tag{5.21}$$

which is the analog of equation 2.5.

5.2.1 Newtonian fluids in laminar flow in pipes

When the fluid is Newtonian and the flow is in the laminar region

$$-\frac{\mathrm{d}u}{\mathrm{d}r} = \frac{\tau}{\mu} \tag{5.22}$$

Substituting in equation 5.18

$$\begin{aligned} \bar{V} &= \frac{D}{2\tau_\mathrm{w}^3}\int_0^{\tau_\mathrm{w}} \frac{\tau^3}{\mu}\mathrm{d}\tau \\ &= \frac{D}{8\mu}\tau_\mathrm{w} \end{aligned} \tag{5.23}$$

or

$$\tau_\mathrm{w} = \frac{8\bar{V}}{D}\mu = \frac{1}{2}\rho_\mathrm{sl}\bar{V}^2 f_\mathrm{sl} \tag{5.24}$$

Re-arranging

$$f_\mathrm{sl} = \frac{16\mu}{D\bar{V}\rho} = \frac{16}{Re} \tag{5.25}$$

5.2.1.1 Velocity profile for Newtonian fluids in laminar motion

When any fluid flows inside a pipe the axial velocity varies strongly with radial position. The shearing action of the fluid against the pipe wall causes the velocity to be low close to the wall and considerably higher close to axis of the pipe. The radial profile in the axial velocity is easy to calculate whenever the fluid is flowing slowly enough to be in laminar motion. The velocity profile is calculated by integration of the differential equation 5.22 using the rheological model for the fluid.

$$\frac{\mathrm{d}u}{\mathrm{d}r} = -\frac{2\tau_\mathrm{W}}{\mu D} r \tag{5.26}$$

with the boundary condition

$$u = 0 \quad \text{at} \quad r = \frac{D}{2} \tag{5.27}$$

The result is

$$u = \frac{\tau_\mathrm{W} D}{4\mu}\left(1 - \left(\frac{2r}{D}\right)^2\right) \tag{5.28}$$

5.2.2 Bingham plastics in laminar flow in pipes

Substituting the rheological model for a Bingham plastic into equation 5.18

$$\begin{aligned}\bar{V} &= \frac{D}{2\tau_\mathrm{W}^3 \mu_\mathrm{B}} \int_{\tau_\mathrm{Y}}^{\tau_\mathrm{W}} \tau^2(\tau - \tau_\mathrm{Y})\mathrm{d}\tau \\ &= \frac{D\tau_\mathrm{W}}{8\mu_\mathrm{B}}\left(1 - \frac{4}{3}\frac{\tau_\mathrm{Y}}{\tau_\mathrm{W}} + \frac{1}{3}\left(\frac{\tau_\mathrm{Y}}{\tau_\mathrm{W}}\right)^4\right)\end{aligned} \tag{5.29}$$

This is called the Buckingham equation and is often written in the form

$$\frac{8\bar{V}\mu_\mathrm{B}}{\tau_\mathrm{y} D}\frac{\tau_\mathrm{Y}}{\tau_\mathrm{W}} = 1 - \frac{4}{3}\frac{\tau_\mathrm{Y}}{\tau_\mathrm{W}} + \frac{1}{3}\left(\frac{\tau_\mathrm{Y}}{\tau_\mathrm{W}}\right)^4 \tag{5.30}$$

or

$$\frac{16}{f_\mathrm{sl} Re_\mathrm{B}} = 1 - \frac{4}{3}\frac{\tau_\mathrm{Y}}{\tau_\mathrm{W}} + \frac{1}{3}\left(\frac{\tau_\mathrm{Y}}{\tau_\mathrm{W}}\right)^4 \tag{5.31}$$

$Re_\mathrm{B} = \dfrac{D\bar{V}\rho_\mathrm{sl}}{\mu_\mathrm{B}}$ is the Reynolds number for the Bingham plastic.

The Hedstrom number $He = (D^2\rho_\mathrm{sl}\tau_\mathrm{Y}/\mu_\mathrm{B}^2)$ is also useful for the characterization of the rheological properties of the Bingham plastic and equation 5.31 can be written in a number alternative ways which are useful for the solution of particular problems.

$$8Re_\mathrm{B}\frac{\tau_\mathrm{Y}}{\tau_\mathrm{W}} = He\left(1 - \frac{4}{3}\frac{\tau_\mathrm{Y}}{\tau_\mathrm{W}} + \frac{1}{3}\left(\frac{\tau_\mathrm{Y}}{\tau_\mathrm{W}}\right)^4\right) \tag{5.32}$$

$$f_\mathrm{sl} = \frac{16}{Re_\mathrm{B}} + \frac{4}{3}f_\mathrm{sl}\frac{\tau_\mathrm{Y}}{\tau_\mathrm{W}} - \frac{1}{3}f_\mathrm{sl}\left(\frac{\tau_\mathrm{Y}}{\tau_\mathrm{W}}\right)^4 \tag{5.33}$$

Substituting for τ_Y in terms of the Hedstrom number and for τ_w from equation 5.20 gives

$$f_{sl} = \frac{16}{Re_B}\left(1 + \frac{1}{6}\frac{He}{Re_B} - \frac{1}{3}\frac{He^4}{f_{sl}^3 Re_B^7}\right) \tag{5.34}$$

Equation 5.34 does not provide an explicit value for f_{sl} but is nevertheless easy to solve for f_{sl} for any specific combination of Re_B and He.

5.2.2.1 Velocity profile for a Bingham plastic in a round pipe

The velocity radial profile of the axial velocity when a Bingham plastic flows steadily in laminar flow in a round pipe can be calculated by integrating the differential equation 5.2 together with equation 5.15

$$\frac{du}{dr} = \frac{\tau_Y}{\mu_B} - \frac{2\tau_w}{\mu_B D}r \tag{5.35}$$

over the region of the pipe where $\tau > \tau_Y$ and

$$\frac{du}{dr} = 0 \tag{5.36}$$

over regions of the pipe where $\tau \leq \tau_Y$.

Since the shear stress in the fluid varies linearly with radius, it is possible to identify these regions explicitly as

$$\tau \leq \tau_Y \quad \text{when} \quad r \leq \frac{\tau_Y}{\tau_w}\frac{D}{2} = r_Y \tag{5.37}$$

r_Y is called the yield radius. If $r_Y > D/2$ no flow occurs and the slurry remains as an immobile solid plug in the pipe. This condition is equivalent to the requirement that $\tau_w > \tau_Y$ for flow to occur and the available pressure gradient must satisfy

$$PDGTF > \frac{4\tau_Y}{D} \tag{5.38}$$

Integrating equations 5.35 and 5.36 from the boundary condition $u = 0$ at $r = D/2$ gives

$$\begin{aligned} u(r) &= \frac{D\tau_w}{4\mu_B}\left(1 - \left(\frac{2r}{D}\right)^2\right) - \frac{D\tau_Y}{2\mu_B}\left(1 - \frac{2r}{D}\right) \quad \text{for} \quad r > \frac{\tau_Y}{\tau_w}\frac{D}{2} \\ &= \frac{D\tau_w}{4\mu_B}\left(1 - \frac{\tau_Y}{\tau_w}\right)^2 \quad \text{for} \quad r \leq \frac{\tau_Y}{\tau_w}\frac{D}{2} \end{aligned} \tag{5.39}$$

The flow pattern consists of a solid plug of undeformed paste of radius τ_Y on the axis surrounded by an annulus of flowing fluid in which the velocity profile is parabolic. This is illustrated in Figure 5.5.

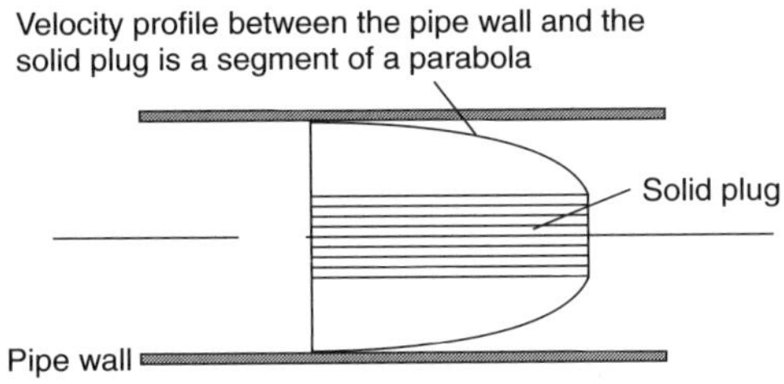

Figure 5.5 Velocity profile when a Bingham plastic flows under laminar conditions in a round pipe

5.2.3 Turbulent flow of Bingham plastics in pipes

5.2.3.1 Laminar-turbulent transition

The flow of a Bingham plastic is defined by two dimensionless groups the Reynolds number Re_B and the Hedstrom number He. The transition from laminar to turbulent flow is defined approximately by the empirical relationship

$$\frac{\dfrac{\tau_Y}{\tau_{wc}}}{\left(1-\dfrac{\tau_Y}{\tau_{wc}}\right)^3} = \frac{He}{16800} \tag{5.40}$$

where τ_{wc} is the wall shear stress at the transition. The Reynolds number at the transition is then given by equation 5.32

$$Re_c = \frac{He}{8}\frac{\tau_{wc}}{\tau_Y}\left(1-\frac{4}{3}\frac{\tau_Y}{\tau_{wc}}+\frac{1}{3}\left(\frac{\tau_Y}{\tau_{wc}}\right)^4\right) \tag{5.41}$$

Illustrative example 5.1

The rheological properties of a dense china clay suspension having density 1600 kg/m^3 can be approximated by the Bingham plastic model with $\tau_Y = 15$ Pa and $\mu_B = 0.15$ Pa s. Calculate the Reynolds number at which the laminar-turbulent transition occurs when this fluid flows in a 10-cm diameter pipe.

Solution

$$He = \frac{D^2 \rho_{sl} \tau_Y}{\mu_B^2} = \frac{0.12^2 \times 1600 \times 15}{0.15^2} = 1.07 \times 10^4$$

Substitution in equation 5.40 gives an equation for the critical value of the wall stress

$$\frac{\dfrac{\tau_Y}{\tau_{wc}}}{\left(1-\dfrac{\tau_Y}{\tau_{wc}}\right)^3} = \frac{1.07 \times 10^4}{16800} = 0.647$$

This equation has solution $(\tau_Y/\tau_{wc}) = 0.261$. Substitution in equation 5.40 gives the Reynolds number at the transition

$$Re_c = \frac{1.07 \times 10^4}{8} \times \frac{1}{0.261}\left(1 - \frac{4}{3} \times 0.261 + \frac{1}{3}(0.261)^4\right) = 3.35 \times 10^3$$

5.2.3.2 Friction factor for Bingham plastics in turbulent flow

An approximate analysis based on Prandtl's mixing length concept is used to develop a relationship between the friction factor and the Reynolds number when a Bingham plastic flows under turbulent conditions in a round pipe.

The average velocity in the pipe is given by equation 5.19

$$\bar{V} = \int_0^1 \xi^2\left(-\frac{\mathrm{d} < u >}{\mathrm{d}\xi}\right)\mathrm{d}\xi \tag{5.42}$$

where $< u >$ is the time-averaged axial velocity. Hanks and Dadia (1971) recommend that the time-averaged velocity gradient be calculated from

$$-\frac{\mathrm{d} < u >}{\mathrm{d}\xi} = \left(\frac{D^{*3}\tau_W}{2\rho_{sl}}\right)^{1/2} G(\xi, \xi_0, R) \quad \text{for} \quad \xi > \xi_0 \tag{5.43}$$

with

$$G(\xi, \xi_0, D^*) = \frac{\xi - \xi_0}{1 + (1 + 0.0648D^{*3}(\xi - \xi_0)(1 - \xi)^2(1 - \exp(-))^{1/2}} \tag{5.44}$$

$$\bar{V} = \left(\frac{D^{*3}\tau_W}{2\rho_{sl}}\right)^{1/2} \int_{\xi_0}^1 \xi^2 G(\xi, \xi_0, D^*)\mathrm{d}\xi \tag{5.45}$$

and the Reynolds number is given by

$$Re_B = \frac{D\bar{V}\rho_{sl}}{\mu_{sl}} = \frac{D^{*3}}{2}\int_{\xi_0}^1 \xi^2 G(\xi, \xi_0, D^*)\mathrm{d}\xi \tag{5.46}$$

D^* is the dimensionless pipe diameter defined by

$$D^* = (Re_B^2 f_{sl})^{1/3} \tag{5.47}$$

The integral in equation 5.46 can be evaluated by numerical quadrature[1] to define the relationship between the Reynolds number and the dimensionless pipe diameter D^* for the Bingham plastic. The friction factor can be readily obtained from $f_{sl} = (D^{*3}/Re_B^2)$ once the average velocity $\bar{V}$, and thus the Reynolds number, have been calculated. f_{sl} is plotted against the Reynolds number in Figure 5.6 together with some experimental data. The smooth

[1]The apparent simplicity of the quadrature is deceptive once the integrand is poorly behaved and strict error control must be implemented for successful evaluation.

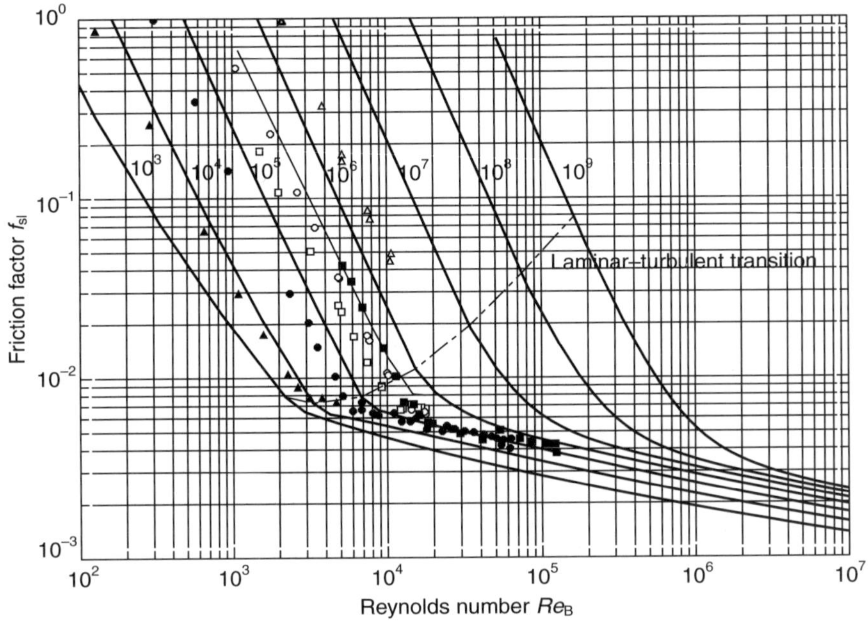

Figure 5.6 Friction factor for Bingham plastics calculated using the Prandtl mixing length model in the transitions and turbulent regions. The Buckingham equation is used in the laminar region. Experimental data from Wilhelm *et al.* (1939) (open symbols) and from Thomas (1960) (filled symbols)

transition from laminar to turbulent flow along the transition locus is a justification for the use of equation 5.40 to define the transition.

Figure 5.6 is not always a convenient aid for the solution of practical problems and the friction factor is presented as a function of the dimensionless variables D^*, V^*, and Q^* in Figures 5.7 to 5.9. These are analogous to the equivalent plots presented in Chapter 2 for Newtonian fluids. These plots are convenient aids for the solution of practical fluid flow problems when the available pressure gradient is known and the pipe diameter, the volumetric flowrate or the flow velocity is known respectively. The use of this type of plot is described in Section 2.3. Under steady flow conditions D^*, Q^* and V^* can be related to the properties of the flow system as shown below.

When the slurry flows at steady state

$$f_{sl} = \frac{D\ PGDTF}{2\rho_{sl}\bar{V}^2} \tag{5.48}$$

and

$$D^{*3} = Re_B^2 f_{sl} = \frac{D^3 \rho_{sl} PGDTF}{2\mu_B^2} = \frac{D\ PGDTF}{2\tau_Y} He = \frac{2\tau_w}{\tau_Y} He \tag{5.49}$$

Thus D^* can be evaluated without having to know the slurry velocity or flowrate.

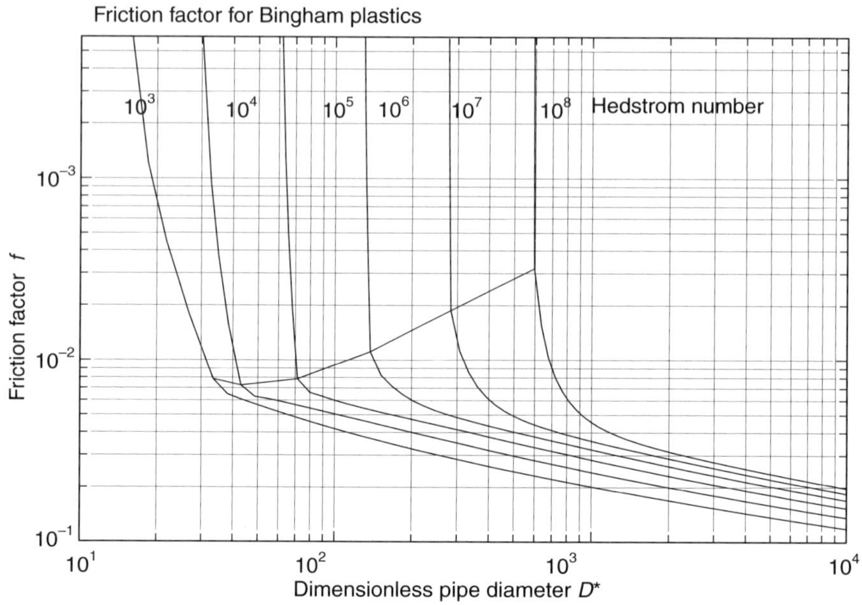

Figure 5.7 Friction factor for Bingham plastics plotted against the dimensionless pipe diameter. Use this plot when the PGDTF and the pipe diameter are known

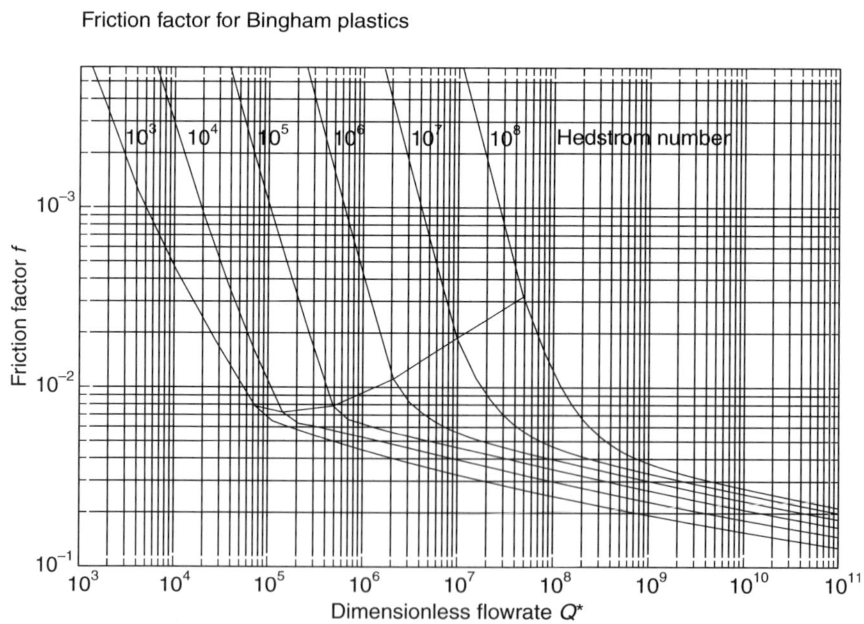

Figure 5.8 Friction factor for Bingham plastics plotted against the dimensionless flowrate. Use this plot when the PGDTF and the desired flowrate are known

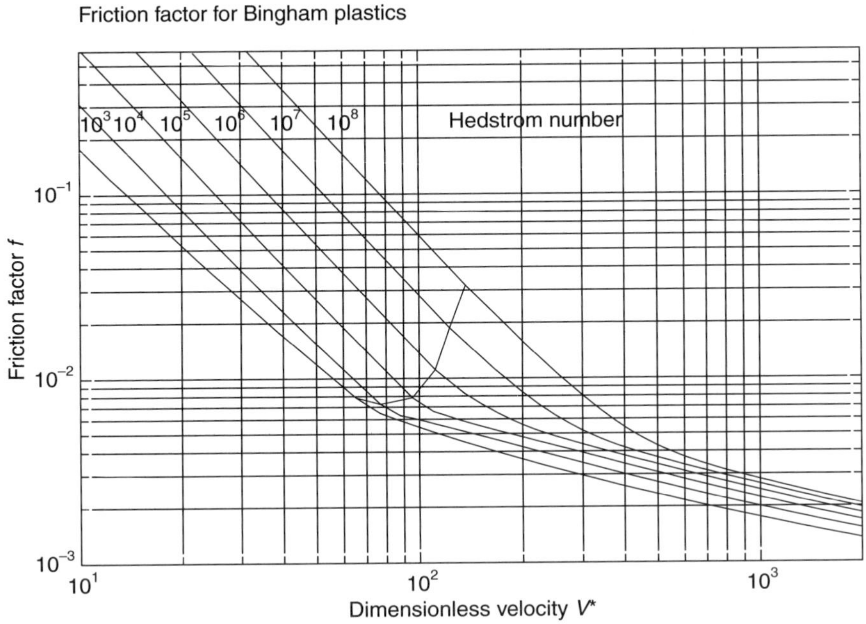

Figure 5.9 Friction factor for Bingham plastics plotted against the dimensionless fluid velocity. Use this plot when the PGDTF and the fluid velocity are known

A minimum value of *PGDTF* is required to initiate flow of a Bingham plastic. The critical value of D^* for flow initiation is obtained when $\tau_W = \tau_Y$ which gives

$$D^*_{\text{init}} = (2He)^{1/3} \tag{5.50}$$

and D^* must be greater than D^*_{init} for flow to take place. The existence of the critical value of D^* is evident from Figure 5.7 where the curve of f_{sl} against D^* approaches the critical value asymptotically at each value of *He*.

When the friction factor is plotted against D^* the curve approaches the critical value as a vertical asymptote as shown in Figure 5.7.

Other dimensionless groups that are useful are

$$Q^{*3} = Re_B^5 f_{\text{sl}} = \frac{32\rho_{\text{sl}}^4 \times PGDTF \times Q^3}{\pi^3 \mu_B^5} \tag{5.51}$$

$$V^{*3} = \frac{Re_B}{f_{\text{sl}}} = \frac{2\rho_{\text{sl}}^2 \bar{V}^3}{\mu_B \times PGDTF} \tag{5.52}$$

The following illustrative examples illustrate the methods.

Illustrative example 5.2

An application requires the slow removal of a dense laterite slurry from a thickener through a 7-cm diameter pipe. Calculate the minimum pressure

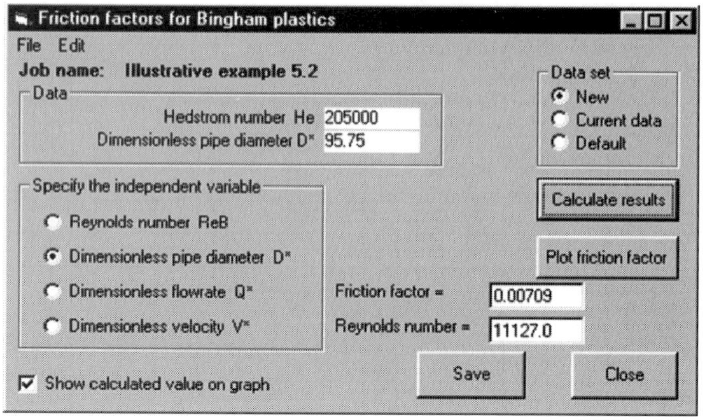

Figure 5.10 Data input screen for the calculation of the friction factor for a Bingham plastic. D^* is the independent variable

gradient required to initiate flow in the pipe. What is the largest pressure gradient for which the flow remains laminar? Calculate the flowrate through the pipe under a pressure gradient of 10 kPa/m (see Figure 5.10).

Bhattacharya *et al.* (1998) have measured the rheological properties of these slurries which behave approximately as Bingham plastics.

Data:

Solids concentration by weight $C_w = 41\%$

Maximum settled concentration by weight $C_{w\,max} = 42.6\%$

d_{50} for the solid particles $= 6.24\,\mu$m.

Density of solid $\rho_s = 3700$ kg/m^3.

pH of the slurry $= 7.1$.

Surface area per unit volume of solids $S_o = 1.79 \times 10^6$ m^2/m^3.

The coefficient of plastic viscosity is related to the concentration in the slurry by

$$\log_{10} \mu_B = 2.414 \frac{C_w}{C_{w\,max}} - 3.6$$

The yield stress is related to the solid concentration of the slurry by

$$\log_{10} \tau_Y = 77.6 \times 10^{-6} \frac{C_w}{C_{w\,max} S_o (d_{50})^{1.227}} - 2.5$$

with d_{50} in μm.

Solution

$$\log_{10} \mu_B = 2.414\frac{41}{42.6} - 3.6 = -1.277$$

$$\mu_B = 52.8 \times 10^{-3} \text{ Pa s}$$

$$\log_{10} \tau_Y = 77.6 \times 10^{-6}\frac{41}{42.6\ 1.79 \times 10^{-6}(6.24)^{1.227}} - 2.5 = 1.913$$

$$\tau_Y = 81.8 \text{ Pa}$$

The density of the slurry is given by

$$\rho_{sl} = \frac{1}{\dfrac{0.41}{3700} + \dfrac{0.59}{1000}} = 1427 \text{ kg/m}^3$$

The pressure gradient at flow initialization is given by equation 5.38

$$PGDTF_{\text{init}} = \frac{4\tau_Y}{D} = \frac{4 \times 81.8}{0.07} = 4.67 \text{ kPa/m}$$

The Hedstrom number for the flow is

$$He = \frac{D^2 \rho_{sl} \tau_Y}{\mu_B^2} = \frac{0.07^2 \times 1427 \times 81.8}{0.0528^2} = 2.05 \times 10^5$$

Substitute this value into equation 5.40 and solve

$$\frac{\tau_Y}{\tau_{wc}} = 0.628$$

The largest pressure gradient for which the flow is laminar is

$$PGDTF_c = \frac{4\tau_{wc}}{D} = \frac{4\tau_Y}{0.628D} = 7.44 \text{ kPa/m}$$

Since the diameter of the pipe and the pressure gradient are known, the solution is most conveniently obtained by calculating D^*

$$D^* = \left(\frac{D^3 \rho_{sl} PGDTF}{2\mu_B^2}\right)^{1/3} = \left(\frac{0.07^3 \times 1427 \times 10 \times 10^3}{2 \times 0.0528^2}\right)^{1/3} = (8.78 \times 10^5)^{1/3} = 95.8$$

Using Figure 5.7, or preferably the Bingham plastic section of the FLUIDS toolbox, to find the applicable value of the friction factor $f_{sl} = 0.00709$ (see Figure 5.11). The Reynolds number is calculated from

$$Re_B = \left(\frac{D^{*3}}{f_{sl}}\right)^{1/2} = \left(\frac{8.78 \times 10^5}{0.00709}\right)^{1/2} = 11127$$

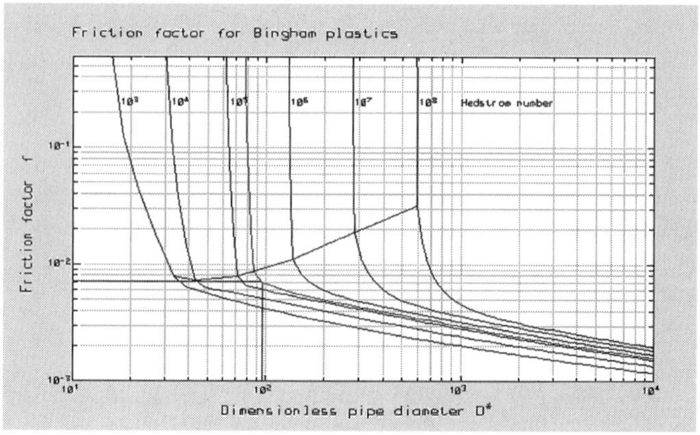

Figure 5.11 Friction factor plot for a Bingham plastic calculated using the FLUIDS toolbox

The velocity is obtained from the Reynolds number

$$\bar{V} = \frac{\mu_B Re_B}{D\rho_{sl}} = \frac{0.0528 \times 11127}{0.07 \times 1427} = 5.88 \text{ m/s}$$

The flowrate is given by

$$Q = \frac{\pi}{4} D^2 \bar{V} = 0.0226 \text{ m}^3/\text{s}$$

Bhattacharya *et al.* (1998) and Huynh *et al.* (2000) have shown that the rheological properties of dense suspensions of this type vary significantly with the pH of the slurry due to the effect on the surface charge on the solid particles. This can affect the flowrate of the slurry and the energy requirements for pumping.

Illustrative example 5.3

What pipe diameter is required to discharge 75 m^3/hr of the slurry defined in example 5.2 under the same pressure gradient.

$$Q = \frac{75}{3600} = 0.0208 \text{ m}^3/\text{s}$$

$$Q^{*3} = \frac{32 \times \rho_{sl}^4 \times PGDTF \times Q^3}{\pi^3 \times \mu_B^5}$$

$$= \frac{32 \times 1427^4 \times 10 \times 10^3 \times 0.0208^3}{\pi^3 \times 0.0528^5} = 9.385 \times 10^{17}$$

$$Q^* = 9.79 \times 10^5$$

The friction factor can be read from Figure 5.8 or more conveniently from the FLUIDS toolbox using an estimated value of the Hedstrom

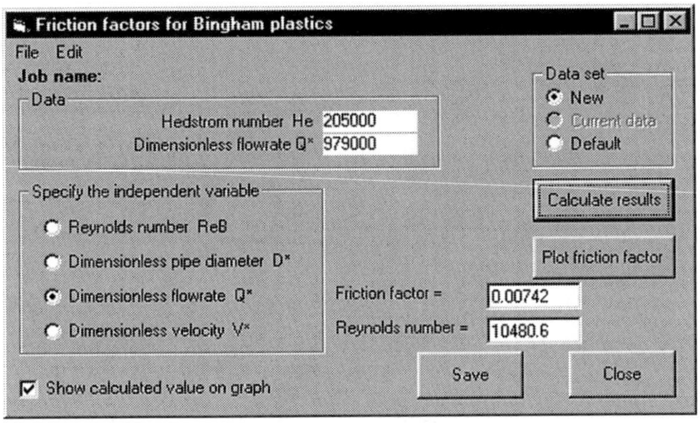

Figure 5.12 Data input screen for the calculation of the friction factor for a Bingham plastic. Q^* is the independent variable

number $He = 205000$. This gives $f_{\text{sl}} = 0.00742$ and $Re_{\text{B}} = 10480.6$ (see Figure 5.12).

$$D = \frac{4Q\rho_{\text{sl}}}{\pi\mu_{\text{B}}Re_{\text{B}}} = \frac{4 \times 0.0208 \times 1427}{\pi \times 0.0528 \times 10480.6} = 6.83 \text{ cm}$$

This estimate for D must now be refined by calculating a new value for He and repeating the calculation.

Equation 5.46 is not suitable for manual calculations and because of the difficult quadrature that is required, is not suitable for spreadsheet calculations. The FLUIDS computational toolbox provides a convenient practical implementation of the method as described in the examples presented above.

Darby (1988) has proposed a simple empirical model that adequately represents the friction factor for a Bingham plastic over the laminar, transition and turbulent regions. In the turbulent region, a Blasius-type equation is used

$$f_{\text{slT}} = 10^a \, Re_{\text{B}}^{-0.193} \tag{5.53}$$

The rheological character of the fluid influences the parameter a

$$a = -1.378\left(1 + 0.146\exp(-2.9 \times 10^{-5} He)\right) \tag{5.54}$$

The complete friction factor curve can be constructed from the laminar and turbulent components using the combination rule

$$f_{\text{sl}} = \left(f_{\text{slL}}^{\beta} + f_{\text{slT}}^{\beta}\right)^{\frac{1}{\beta}} \tag{5.55}$$

where f_{slL} is the friction factor for laminar flow given by equation 5.34 and the combination parameter varies with the Reynolds number as follows.

$$\beta = 1.7 + \frac{40000}{Re_{\text{B}}} \tag{5.56}$$

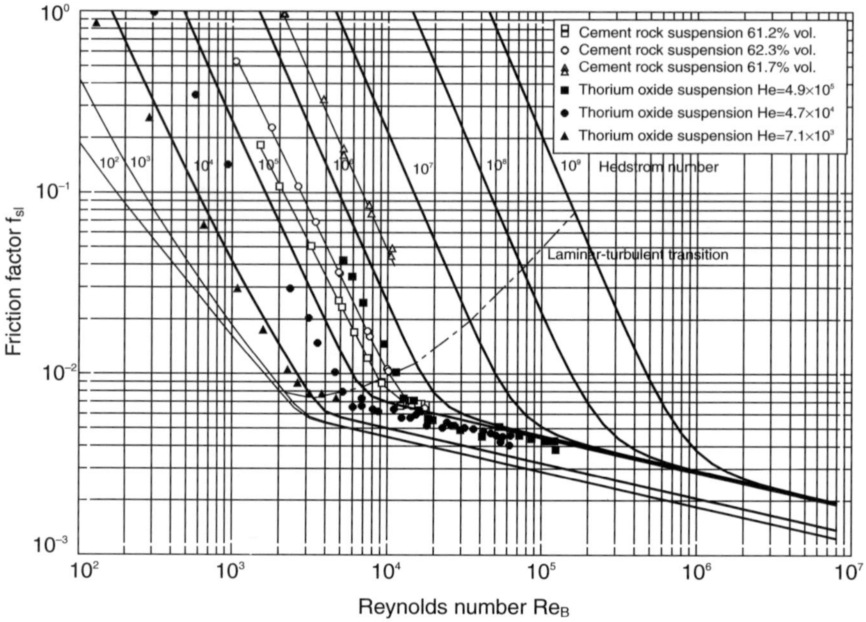

Figure 5.13 Friction factor plot for a Bingham plastic calculated using Darby's empirical model. Experimental data from Wilhelm *et al.* (1939) (open symbols) and from Thomas (1960) (filled symbols)

The friction factor calculated using this model is shown in Figure 5.13. The locus that represents the transition from laminar to turbulent flow that is shown in Figure 5.13 is defined by equation 5.40. Equation 5.55 does not require a specific transition point since it is based on a seamless smooth transition from the laminar to the turbulent regime of flow.

Darby's equation is convenient when manual calculation methods must be used.

5.2.4 Yield – pseudo plastic fluids in laminar flow in pipes

The Herschel–Bulkley model describes the behavior of pseudo plastic fluids that exhibit a yield stress. The flowrate of these fluids in laminar flow in a round pipe can be calculate by substituting the model equation 5.8 into equation 5.18

$$\bar{V} = \frac{D}{2\,K_{\mathrm{H}}^{1/n}\,\tau_{\mathrm{w}}^{3}} \int_{\tau_{\mathrm{h}}}^{\tau_{\mathrm{w}}} \tau^{2}(\tau - \tau_{\mathrm{H}})^{1/n}\,\mathrm{d}\tau \tag{5.57}$$

A change of variable to

$$u = \tau - \tau_{\mathrm{H}} \tag{5.58}$$

leads to a simple integral

$$\bar{V} = \frac{D}{2\,K_H^{1/n}\,\tau_w^3}\int_0^{\tau_w - \tau_H}\left(u^{2+1/n} + 2\,\tau_H\,u^{1+1/n} + \tau_H^2\,u^{1/n}\right)du$$
$$= \frac{D}{2\,K_H^{1/n}\,\tau_w^3}\left(\frac{4n}{3n+1}\left(1-\frac{\tau_H}{\tau_w}\right)^{\frac{3n+1}{n}} + \frac{8n}{2n+1}\frac{\tau_h}{\tau_w}\left(1-\frac{\tau_H}{\tau_w}\right)^{\frac{2n+1}{n}}\right. \tag{5.59}$$
$$\left. + \frac{4n}{n+1}\left(\frac{\tau_H}{\tau_w}\right)^2\left(1-\frac{\tau_H}{\tau_w}\right)^{\frac{n+1}{n}}\right)$$

5.2.5 Power-law fluids in laminar flow in pipes

The relationship between the shear stress and the velocity gradient in a power-law fluid is given by equation 5.7

$$-\frac{du}{dr} = \left(\frac{\tau}{K}\right)^{1/n} \tag{5.60}$$

In this model the parameter K has the dimensions Pa s^n and the index n is dimensionless.

Substituting this into equation 5.18 gives

$$\bar{V} = \frac{D}{2\,\tau_w^3\,K^{1/n}}\int_0^{\tau_w}\tau^{2+1/n}\,d\tau$$
$$= \frac{nD}{2(3n+1)}\left(\frac{\tau_w}{K}\right)^{1/n} \tag{5.61}$$

Equation 5.61 is usually written in the form

$$\tau_w = K\left(\frac{3n+1}{4n}\right)^n\left(\frac{8\bar{V}}{D}\right)^n = \frac{1}{2}\rho_{sl}\,\bar{V}^2 f_{PL} \tag{5.62}$$

A modified Reynolds number is defined for a power-law fluid as

$$Re_{PL} = \frac{D\bar{V}\,\rho_{sl}}{K}\left(\frac{4n}{3n+1}\right)^n\left(\frac{8\bar{V}}{D}\right)^{1-n} \tag{5.63}$$

then equation 5.62 gives

$$f_{PL} = \frac{16}{Re_{PL}} \tag{5.64}$$

for laminar flow of a power-law fluid. Equation 5.64 can be compared with equation 5.25 which is the equivalent expression for Newtonian fluids.

The group

$$\mu_{PL} = K\,8^{n-1}\left(\frac{3n+1}{4n}\right)^n \tag{5.65}$$

is called the generalized coefficient of viscosity for power law fluids and Re_{PL} is sometimes written as

$$Re_{PL} = \frac{D\bar{V}\rho_{sl}}{\mu_{PL}}\left(\frac{\bar{V}}{D}\right)^{1-n} \tag{5.66}$$

Equation 5.62 can be used to determine the value of n experimentally using a tube viscometer. τ_w plotted against the group $8\bar{V}/D$ on logarithmic coordinates will produce a straight line of slope n from which n and K can be calculated.

Illustrative example 5.3

The data shown in Table 5.2 were obtained during a test of a chalk slurry in a laboratory pipeline having a diameter of 15 mm. Show that the data are consistent with a power-law model for the slurry and evaluate the flow parameters n, K and μ_{PL}.

Table 5.2 Test flow data for a chalk slurry

Pressure gradient Pa/m	24.1	48.9	115.1	185.9
Flowrate ml/s	1.20	3.53	13.3	27.8

The data can be checked against equation 5.62 in the form

$$\tau_w = 8^{1-n}\,\mu_{PL}\left(\frac{8\bar{V}}{D}\right)^n$$

The data can be converted using

$$\tau_w = \frac{D}{4} \times PGDTF$$

and

$$\frac{8\bar{V}}{D} = \frac{32Q}{\pi D^3}$$

The converted data are plotted in Figure 5.14 which shows the straight line relationship confirming that the slurry can be modeled as a power-law fluid. The value of n is calculated from the slope of the line

$$n = \frac{log\left(\frac{185.0}{24.1}\right)}{log\left(\frac{27.8}{1.20}\right)} = 0.65$$

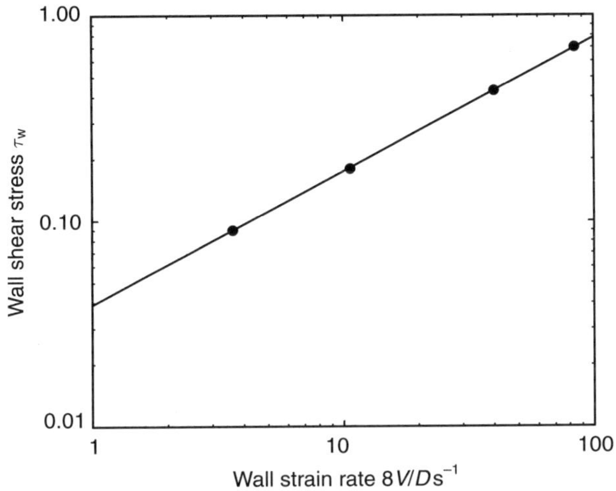

Figure 5.14 Experimental data for illustrative example 5.3

When $Q = 27.8\,\text{ml/s}$

$$\bar{V} = \frac{4 \times 27.8 \times 10^{-6}}{\pi \times 0.015^2} = 0.157\,\text{m/s}$$

$$\frac{8\bar{V}}{D} = \frac{8 \times 0.157}{0.015} = 83.9\ s^{-1}$$

$$\tau_w = \frac{0.015 \times 185.9}{4} = 0.697\,\text{Pa}$$

$$\mu_{PL} = \frac{0.697}{8^{0.35} \times 83.9^{0.65}}\ \text{Pa s}^{0.65}$$

Check for laminar flow during the experiment.

$$Re_{PL} = \frac{0.015 \times 0.157 \times 1200}{0.0189}\left(\frac{0.157}{0.015}\right)^{0.35} = 340$$

The flow is laminar over the range of the data confirming that equation 5.62 is applicable.

5.3 Power-law fluids in turbulent flow in pipes

5.3.1 Dense slurries

The power-law model provides an alternative to the Bingham plastic model for concentrated non-settling slurries. The power-law model also describes the flow behavior of many polymer solutions although the relationship between the friction factor and the Reynolds number differs from that required to describe the behavior of concentrated slurries. The friction factor is correlated against the power-law Reynolds number for both fluid types but different equations are required for each. An empirical friction factor equation has been found to correlate the experimental data for turbulent flow of concentrated slurries that behave as power-law fluids.

$$f_{PL} = \frac{0.25E}{(Re_{PL})^m}\phi^{\frac{1000}{Re_{PL}}} \tag{5.67}$$

Where f_{PL} is called the power-law friction factor and the parameters in the equation are related to the power-law index n through the following equations.

$$\begin{aligned} E &= 0.0089\exp(3.57\,n^2) \\ m &= 0.314\,n^{2.3} - 0.064 \\ \phi &= \exp\left(0.572\frac{1-n^{4.2}}{n^{0.435}}\right) \end{aligned} \tag{5.68}$$

Equation 5.67 was developed by Kemblowski and Kolodziejski (1973) and we refer to it as the Kemblowski–Kolodziejski equation.

The power-law index n is less than 1 for pseudoplastic fluids typical values range from 0.2 to 0.9. When $n = 1$ the fluid will exhibit Newtonian behavior and equations 5.68 give $E = 0.316$, $m = 0.25$ and $\varphi = 1$. When these values are substituted into equation 5.67

$$Re_{PL} = Re \quad \text{and} \quad f_{PL} = \frac{0.079}{Re^{0.25}} \tag{5.69}$$

which is equivalent to the Blasius equation for Newtonian fluids.

Equation 5.67 provides a convenient correlation for turbulent flow of power-law fluids and is a useful tool for the solution of practical fluid flow problems. The friction factor calculated from equation 5.67 is plotted against the power-law Reynolds number in Figure 5.15. Data determined in experiments by Metzner and Reed (1955) and Kemblowski and Kolodziejski (1973) are shown in Figure 5.15 for comparison with equation 5.67. It can be seen from the graph that the friction factor calculated from equation 5.67 is lower than that for a Newtonian fluids in a transition region between laminar and fully turbulent flow. The friction factor is calculated from the Colebrook equation for Newtonian fluids when the flow is fully turbulent and the Kemblowski–Kolodziejski equation applies only until its graph intersects the Newtonian fluid line after which the fluid behavior is described by the curve for Newtonian fluids as shown in Figure 5.15. The friction factor for power-law fluids under laminar flow conditions in a smooth pipe calculated from equation 5.64 is shown as the straight line in the region $Re_{PL} < 2000$.

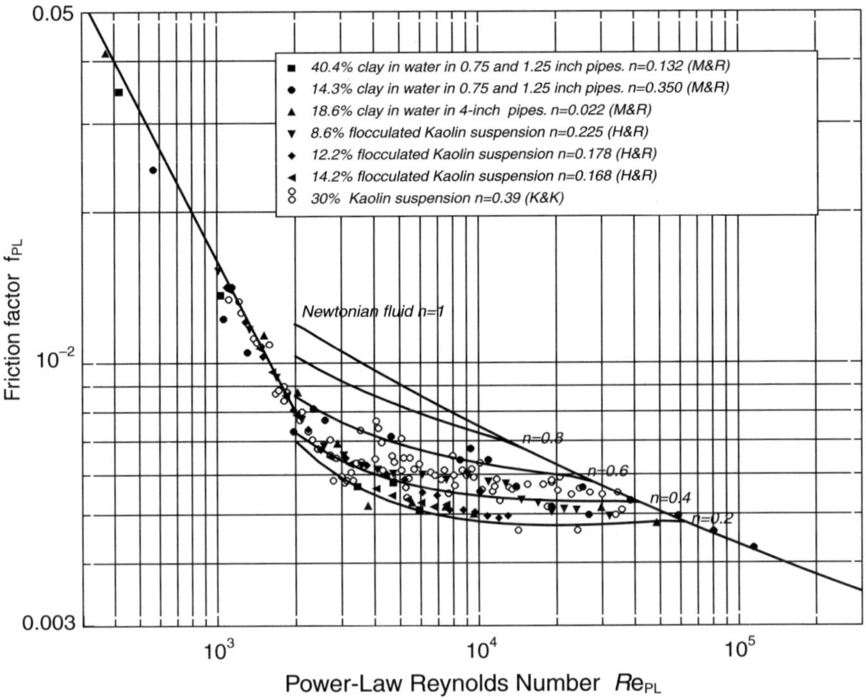

Figure 5.15 Friction factor for a power-law fluid calculated from equation 5.67. Data are from Metzner and Reed (1955), Heywood and Richardson (1978) and from Kemblowski and Kolodziejski (1973) open symbols

---

Illustrative example 5.4

Calculate the pressure gradient due to friction along a 5.7-cm diameter pipe when the chalk slurry described in illustrative example 5.3 flows at a rate of $2.23 \times 10^{-3}\,\text{m}^3/\text{s}$ (see Figure 5.16).

$$\bar{V} = \frac{4Q}{\pi D^2} = \frac{4 \times 2.23 \times 10^{-3}}{\pi 0.057^2} = 0.874\,\text{m/s}$$

$$Re_{\text{PL}} = \frac{0.057 \times 0.874 \times 1200}{0.0189}\left(\frac{0.874}{0.057}\right)^{0.35} = 8.22 \times 10^{-3}$$

Use Figure 5.15, equation 5.67 or the FLUIDS toolbox to get the friction factor.

$$E = 0.0089\exp(3.57 \times 0.65^2) + 0.0402$$

$$m = 0.314 \times 0.65^{2.3} - 0.064 = 0.0526$$

$$\varphi = \exp\left(0.572 \times \frac{1 - 0.65^{4.2}}{0.65^{0.435}}\right) = 1.763$$

$$f_{\text{PL}} = \frac{0.25 \times 0.0402}{(8.22 \times 10^3)^{0.0526}} 1.763^{1/8.22} = 0.00671$$

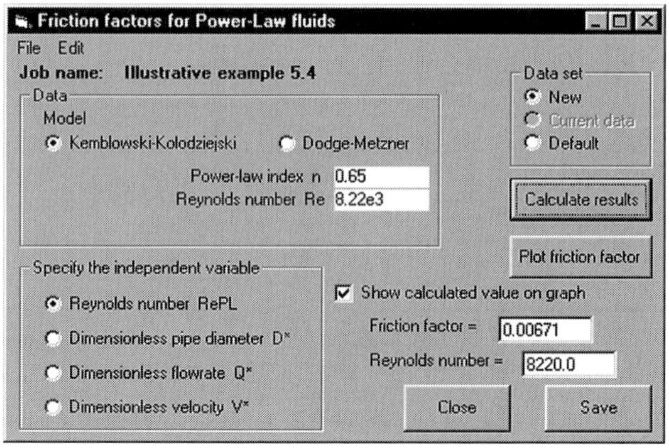

Figure 5.16 Data input screen for calculation of the friction factor for a power-law fluid. The Kemblowski–Kolodziejski model is selected

Use equation 5.21 to get the pressure gradient due to friction

$$PGDTF = \frac{2 \times 1200 \times 0.874^2 \times 0.00671}{0.057} = 215.8 \text{ Pa/m}$$

Figure 5.15 can be used directly for a power-law fluid of known flow index n whenever the pipe diameter and average fluid velocity are known. Under these conditions, Re_{PL} can be evaluated using equation 5.63 and the friction factor read from Figure 5.15 or calculated from equation 5.67. When the average slurry velocity is not known and must be calculated, Figure 5.15 is not convenient to use because the Reynolds number cannot be calculated and alternative representations of the relationship between the friction factor f and the Reynolds number Re_{PL} are much more convenient. The average fluid velocity achieved when the pressure gradient and the pipe diameter are known can be calculated for the flow of power-law fluids using the dimensionless pipe diameter which is defined by

$$D^{*^3} = Re_{PL}^2 f_{PL}^{2-n} \tag{5.70}$$

The friction factor for a dense slurry that behaves as a power-law fluid, calculated using the Kemblowski–Kolodziejski equation is plotted against the dimensionless velocity D^* in Figure 5.17.

When the flow is laminar

$$Re_{PL} = \frac{16}{f_{PL}} \tag{5.71}$$

so that

$$f_{PL} = \left(\frac{16^2}{D^{*3}}\right)^{1/n} \tag{5.72}$$

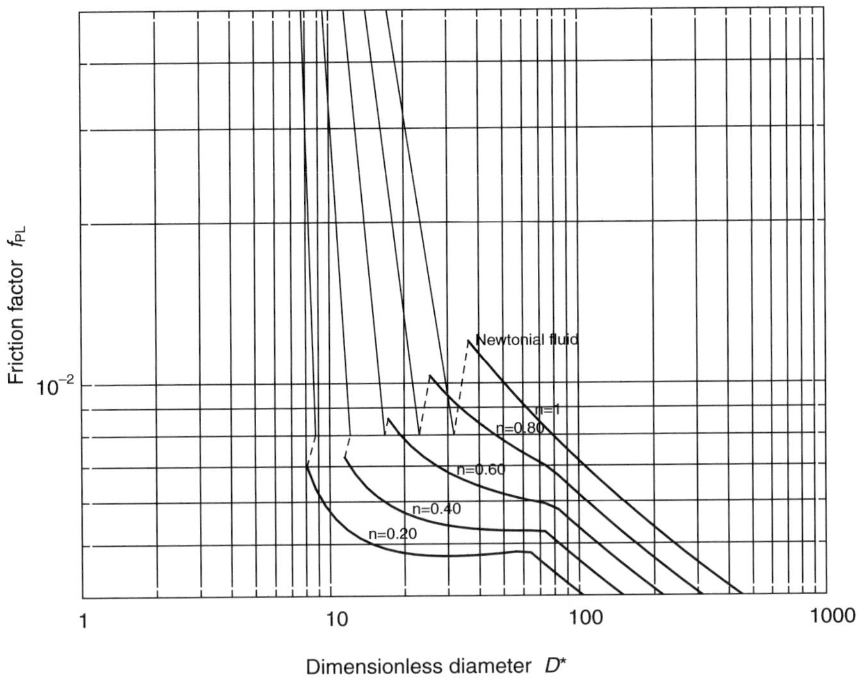

Figure 5.17 Friction factor for a power-law fluid plotted against the dimensionless pipe diameter using the Kemblowski–Kolodziejski equation. Use this plot when the pipe diameter and the *PGDTF* are known.

Thus the laminar portion of the graph is a straight line with slope that depends on the rheological constant n as shown in Figure 5.17.

For a power-law fluid flowing steadily in a round pipe

$$PGDTF = \frac{2\,\rho_{sl}\,\bar{V}^2 f_{PL}}{D} \tag{5.73}$$

and D^* can be written in terms of the physical properties of the fluid, the pipe diameter and the pressure gradient using equation 5.70

$$D^{*3} = \frac{D^{2+n} \times PGDTF^{2-n} \times \rho_{sl}^{n}}{2^{2-n}\,\mu_{sl}^{2}} \tag{5.74}$$

which is independent of $\bar{V}$. This definition of the dimensionless pipe diameter can be compared to that for Newtonian fluids given in Chapter 2.

D^* can be computed for steady flow without having to know $\bar{V}$ and f_{PL} can be read from Figure 5.17. Once f_{PL} is known, the power-law Reynolds number can be calculated from

$$Re_{PL} = \left(\frac{D^{*3}}{f_{PL}^{2-n}}\right)^{\frac{1}{2}} \tag{5.75}$$

When the diameter of the pipe must be selected to achieve a given flowrate under a given pressure driving force, the friction factor should be plotted against the dimensionless flow rate

$$Q^{*3} = Re_{\mathrm{PL}}^{5} f_{\mathrm{sl}}^{4-3n} \tag{5.76}$$

The friction factor from Figure 5.15 is plotted against Q^* in Figure 5.18.

When the fluid flows steadily in a round pipe, f_{PL} can be substituted from equation 5.73 into equation 5.76 to give an expression for Q^* in terms of the physical properties of the fluid and the volumetric flowrate through the pipe and the pressure gradient.

$$Q^{*3} = \frac{2^{5n}\,\rho_{\mathrm{sl}}^{1+3n}\,PGDTF^{4-3n}\,Q^{2+n}}{\pi^{2+n}\,\mu_{\mathrm{PL}}^{5}} \tag{5.77}$$

which is independent of D and the velocity of the fluid.

Q^* can be computed without having to know the pipe diameter. The friction factor can be read from Figure 5.18. The Reynolds number can be recovered from

$$Re_{\mathrm{PL}} = \left(\frac{Q^{*3}}{f_{\mathrm{PL}}^{4-3n}}\right)^{\frac{1}{5}} \tag{5.78}$$

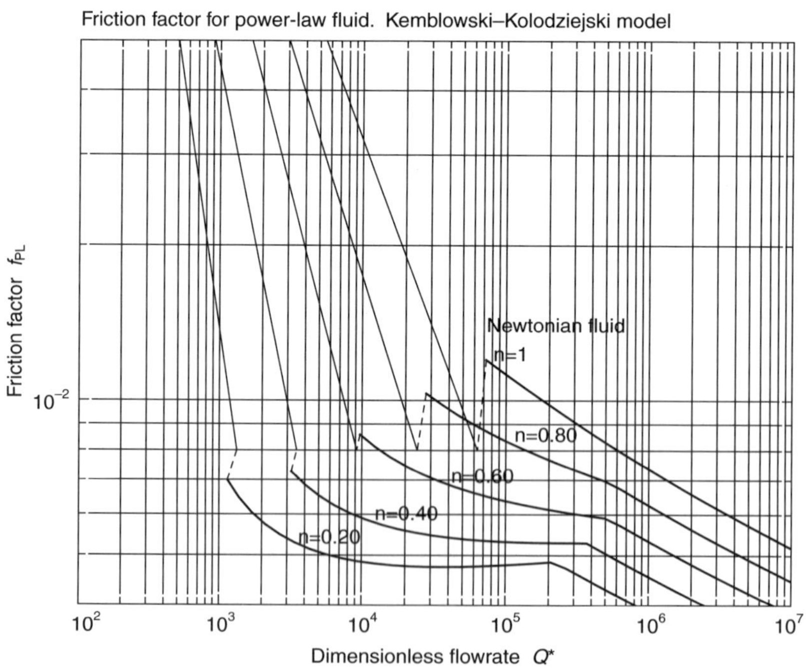

Figure 5.18 Friction factor for a power-law fluid plotted against the dimensionless flowrate. Use this chart when the flowrate and the *PGDTF* are known

The required pipe diameter can be calculated from the Reynolds number

$$Re_{\mathrm{PL}} = \frac{4^{2-n}\,\rho_{\mathrm{sl}}\,Q^{2-n}}{\pi^{2-n}\,\mu_{\mathrm{PL}}\,D^{4-3n}} \tag{5.79}$$

Illustrative example 5.5

Calculate the pipe diameter that is required to pump the slurry of illustrative example 5.3 at a rate of $75\,\mathrm{m}^3/\mathrm{hr}$ under a pressure gradient of 1.2 KPa/m (see Figure 5.19).

$$Q = \frac{75}{3600} = 0.0208\,m^3/s$$

$$Q^{*3} = \frac{2^{3.25} \times 1200^{2.95} \times (1.2 \times 10^3)^{2.05} \times 0.0208^{2.675}}{\pi^{2.65} \times 0.0189^5} = 1.65 \times 10^{19}$$

$$Q^* = 2.546 \times 10^6$$

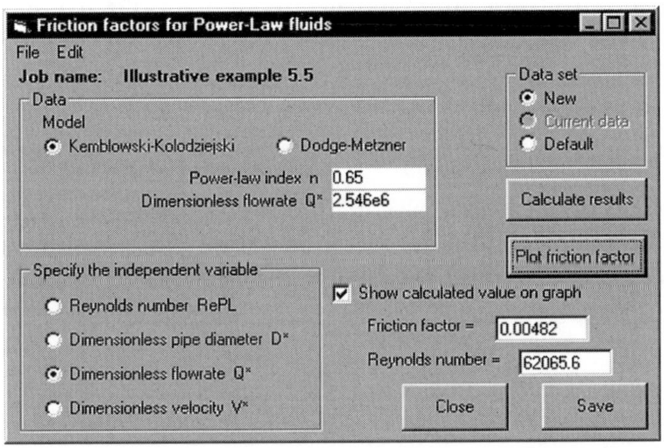

Figure 5.19 Friction factor for a power-law fluid calculated using the FLUIDS toolbox

Use Figure 5.18 or the FLUIDS toolbox to get the friction factor

$$f_{\mathrm{PL}} = 0.00482$$

$$Re_{\mathrm{PL}} = \left(\frac{Q^{*3}}{f_{\mathrm{PL}}^{2.05}}\right)^{\frac{1}{5}} = \left(\frac{1.65 \times 10^{19}}{0.00482^{2.05}}\right) = 6.21 \times 10^4$$

$$= \frac{4^{1.35} \times 1200 \times 0.0208^{1.35}}{\pi^{1.35} \times 0.0189 \times D^{2.05}} = \frac{471.8}{D^{2.05}}$$

Solve for D to get $D = 9.25$ cm (see Figure 5.20).

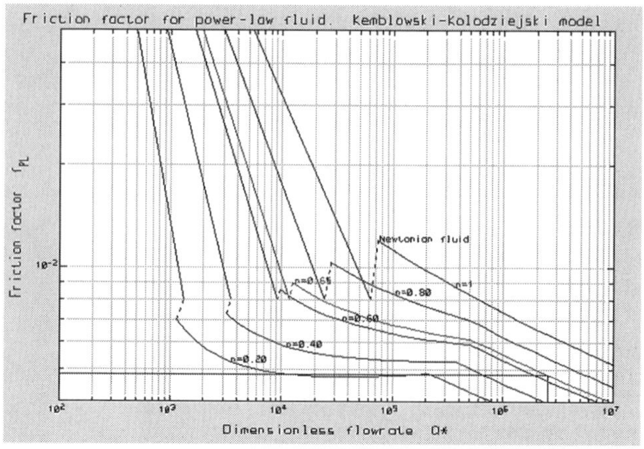

Figure 5.20 Friction factor plot for a power-law fluid calculated using the FLUIDS toolbox

A similar procedure can be used to calculate the required pipe diameter when a dense power-law slurry must flow at a specific average velocity $\bar{V}$ under a prespecified pressure gradient. In this case a dimensionless velocity is defined by

$$V^{*3} = \frac{Re_{PL}}{f_{PL}^{n}} \tag{5.80}$$

The power-law friction factor is plotted against V^* in Figure 5.21.

Under steady flow conditions V^* can be written in terms of the properties of the fluid, the velocity and the pressure gradient by substituting for f_{PL} from equation 5.73.

$$V^{*3} = \frac{2^n \rho_{sl}^{n+1} \bar{V}^{2+n}}{\mu_{PL}^{n} PGDTF^{n}} \tag{5.81}$$

V^* can be calculated and the value of f_{PL} read from Figure 5.21. The Reynolds number is recovered from

$$Re_{PL} = \bar{V}^{*3} f_{PL}^{n} \tag{5.82}$$

The pipe diameter is obtained from

$$Re_{PL} = \frac{D^n \bar{V}^{2-n} \rho_{sl}}{\mu_{PL}} \tag{5.83}$$

5.3.2 Flow of polymer solutions in pipes

An alternative model for the friction factor for power-law fluids is usually applied to polymer solutions and also some slurries. The friction

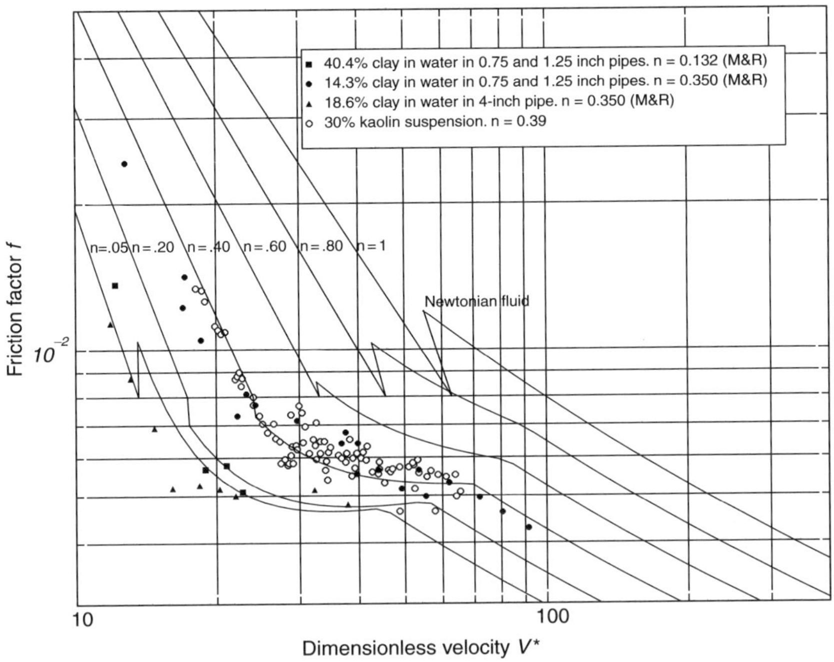

Figure 5.21 Friction factor for a power-law fluid plotted against the dimensionless velocity. Use this chart when the desired fluid velocity and the PGDTF are known

factor in the turbulent region is related to the power-law Reynolds number by

$$\frac{1}{\sqrt{f_{PL}}} = \frac{1.74}{n^{0.75}} \ln\left(f_{PL}^{(2-n)/2}\right) - \frac{0.4}{n^{1.2}} \tag{5.84}$$

which can be compared with the equivalent equation 2.14 for Newtonian fluids. Equation 5.84 is referred to as the Dodge–Metzner equation (1959). This equation is shown in graphical form in Figure 5.22 where it is compared to some experimental data.

The friction factor from the Dodge–Metzner equation can be plotted against the dimensionless variables D^*, Q^*, and V^* to facility the solution of practical problems. These graphs are not shown here but are available in the FLUIDS computational toolbox. The student is encouraged to explore these graphs. They can be found under the power-law section in the toolbox and selecting the Dodge–Metzner option in the data input screen that is shown in Figure 5.19.

5.4 Shear-thinning fluids with Newtonian limit

Some shear-thinning fluids behave as Newtonian fluids at high strain rates while they exhibit power-law behavior at low strain rates. Equation 5.10 provides a model for this type of behavior. Figure 5.3 shows some experimental

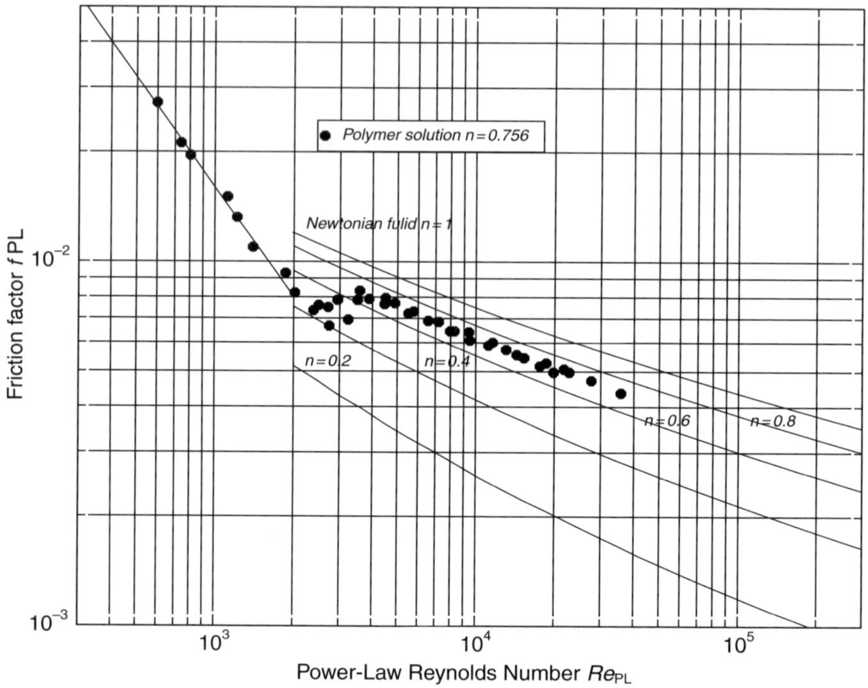

Figure 5.22 Friction factors for power-law fluids calculated using the Dodge–Metzner equation. Experimental data from Metzner and Reed (1955)

data obtained with titanium dioxide slurries. These data show clearly that the log–log plots asymptote to straight lines at both ends of the strain rate range. The asymptote at low strain rate has a slope of n and the asymptote at high strain rate has slope of unity as required by equation 5.10. Newtonian behavior is characterized by a straight line of slope 1 in this coordinate system.

If the strain rate $-du/dr$ is represented by the symbol γ, equation 5.10 becomes

$$\tau = \mu_\infty \gamma + K_S \gamma^n \tag{5.85}$$

When a fluid with these rheological characteristics flows steadily in a pipe of circular cross-section, the average velocity of flow can be related to the pressure gradient. Starting with equation 5.18

$$\bar{V} = \frac{D}{2\tau_w^3}\int_0^{\tau_w} \tau^2 \gamma d\tau \tag{5.86}$$

Substitution of τ from equation 5.85 gives

$$\bar{V} = \frac{D}{2\tau_w^3}\int_0^{\gamma_w} (\mu_\infty \gamma + K_S \gamma^n)^2 \gamma(\mu_\infty + nK_S \gamma^{n-1})d\gamma \tag{5.87}$$

After expansion of the integrand this yields

$$\frac{8\bar{V}}{D} = \frac{4\dot{\gamma}_w}{(1+X)^3}\left(\frac{1}{4} + \frac{n+2}{n+3}X + \frac{2n+1}{2n+2}X^2 + \frac{n}{3n+1}X^3\right) = \dot{\gamma}_w\, G(n, X) \quad (5.88)$$

where

$$X = \frac{K_S}{\mu_\infty}\dot{\gamma}_w^{n-1} \quad (5.89)$$

and

$$G(n, X) = \frac{4}{(1+X)^3}\left(\frac{1}{4} + \frac{n+2}{n+3}X + \frac{2n+1}{2n+2}X^2 + \frac{n}{3n+1}X^3\right) \quad (5.90)$$

both of which are dimensionless. Equation 5.88 defines an implicit relationship between $\dot{\gamma}_w$ and $8\bar{V}/D$ which cannot be solved in closed form. It is convenient to introduce the dimensionless number

$$H = \frac{K_S}{\mu_\infty}\left(\frac{8\bar{V}}{D}\right)^{n-1} \quad (5.91)$$

and equation 5.88 becomes

$$H = XG(n, X)^{n-1} \quad (5.92)$$

The shear stress at the pipe wall is given by

$$\begin{aligned}\tau_w &= \mu_\infty \dot{\gamma}_w + K_S \dot{\gamma}_w^n \\ &= \mu_\infty (1+X)\dot{\gamma}_w \\ &= \mu_\infty (1+X)\frac{8\bar{V}}{D}\frac{1}{G(n, X)}\end{aligned} \quad (5.93)$$

This leads to a relationship between the friction factor and the Reynolds number

$$f_{sl} = \frac{2\tau_w}{\rho_{sl}\bar{V}^2} = \frac{16}{Re_\infty}\frac{(1+X)}{G(n, X)} \quad (5.94)$$

where the Reynolds number is defined by

$$Re_\infty = \frac{D\bar{V}\rho_{sl}}{\mu_\infty} \quad (5.95)$$

The relationship between f_{sl} and Re_∞ depends on the rheological properties of the particular slurry under consideration and on slurry velocity and the diameter of the pipe.

When the flow is turbulent the fluid behaves as a Newtonian fluid of viscosity μ_∞ and the friction factor is given by equation 2.16

$$f_{sl} = 0.079\, Re_\infty^{-0.25} \quad (5.96)$$

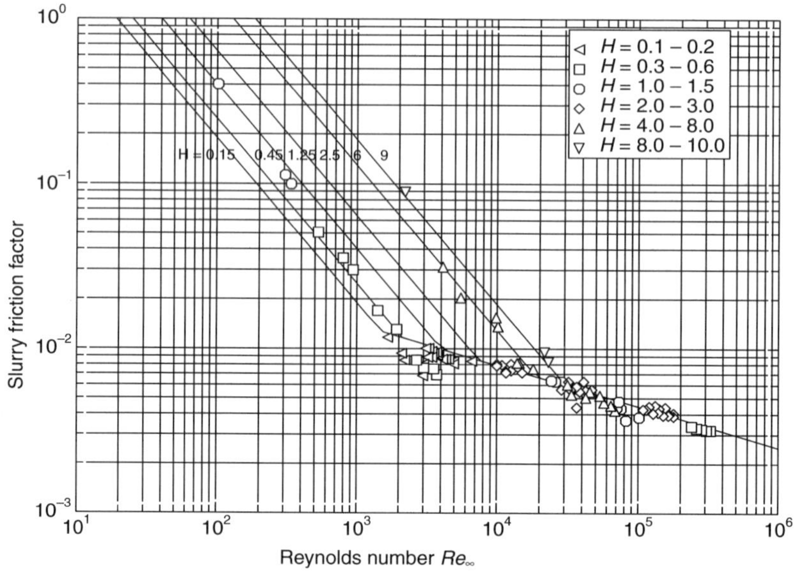

Figure 5.23 Friction factor for a Sisko fluid with flow index $n = 0.249$. Experimental data from Ma (1987) for a 3.6 per cent (vol) laterite slurry with parameters $\mu_\infty = 0.00145$ Pa s, $K_s = 0.396$ Pa s^n and $n = 0.249$ is shown for comparison

Equations 5.94 and 5.96 show that the friction factor vs Reynolds number plot is a series of straight lines in the laminar flow region and a single straight line in the turbulent region when plotted in the log–log coordinate system.

The friction factor is plotted against Re_∞ in Figure 5.23. Experimental data for a 3.6 per cent laterite slurry are shown in this figure and the flow behavior of the slurry is described well by the Sisko model. Because this model has three parameters it is not possible to construct a single curve of the type shown in Figure 5.23 that will apply to all fluids that behave according to the Sisko model. A separate curve is required for each value of the parameter n. It would be impractical to prepare separate figures for all possible values of n and the reader is referred to the FLUIDS toolbox to generate the curve for whatever value of n is required.

Illustrative example 5.6

Calculate the friction factor when a Sisko fluid with index $n = 0.25$ flows in a round pipe. Consider the case when $Re_\infty = 10^3$ and $H = 2.5$.

Solution

Solve equation 5.92

$$\begin{aligned} 0.25 &= XG(n, X)^{n-1} \\ &= \frac{4X}{(1+X)^3}\left(0.25 + 0.6923X + 0.60\,X^2 + 0.1429\,X^3\right) \end{aligned}$$

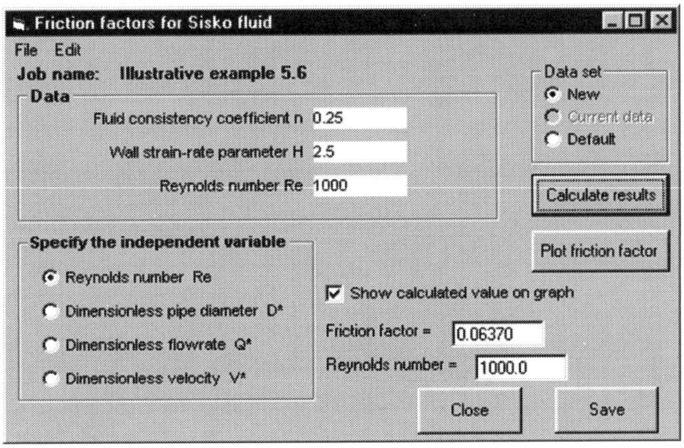

Figure 5.24 Data input screen for calculation of friction factor of a Sisko fluid

to get $X = 2.043$. (This requires the use of a non-linear equation solver)

$$G(0.25, 2.043) = 0.7643$$

$$f_{\text{sl}} = \frac{16}{Re_\infty}\frac{1+X}{G(n,X)} = \frac{16}{10^3}\frac{1+2.043}{0.7643} = 6.37 \times 10^{-2}$$

This can be checked on Figure 5.23.

The friction factor for Sisko fluids can be conveniently obtained by using the Sisko fluid section in the FLUIDS computational toolbox as shown in Figure 5.24.

The friction factor for the Sisko fluid can be plotted against the variables D^*, Q^* and V^* to facilitate the solution of problems that require the calculation of the flowrate or required pipe diameter when the available pressure gradient is known.

$$D^* = \left(Re_\infty^2 f_{\text{sl}}\right)^{1/3} \tag{5.97}$$

Under steady flow conditions

$$f_{\text{sl}} = \frac{D}{2\,\rho_{\text{sl}}\,\bar{V}^2} PGDTF \tag{5.98}$$

Substituting for Re_∞ and f_{sl} into equation 5.97

$$D^* = \left(\frac{\rho_{\text{sl}}\,PGDTF}{2\,\mu_\infty^2}\right)^{1/3} D \tag{5.99}$$

The friction factor is plotted against D^* in Figure 5.25. Like the friction factor graph from which it is generated, this plot is also a series of straight lines as can be seen by substituting equations 5.94 and 5.96 into the definition of D^*.

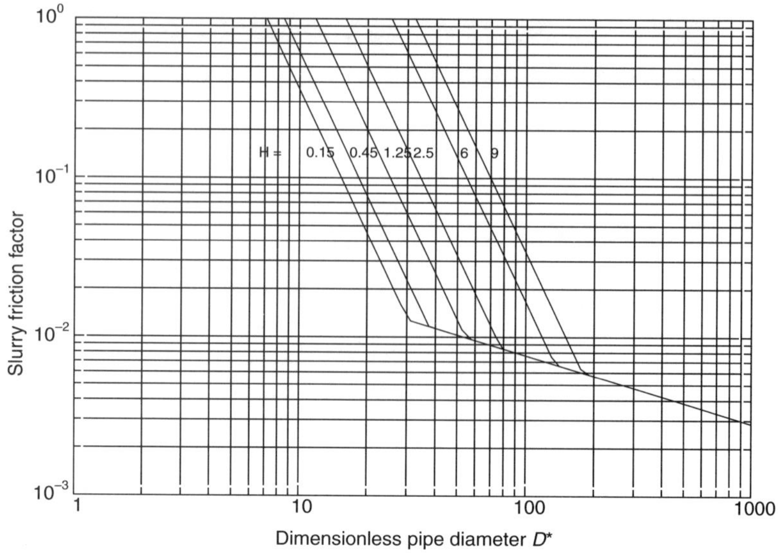

Figure 5.25 Friction factor for a Sisko fluid with flow index $n = 0.249$. Use this plot when the pipe diameter and the available pressure gradient are known

In the laminar regime

$$f_{sl} = \left(16\frac{1+X}{G(n, X)}\right)^2 \frac{1}{D^{*3}} \tag{5.100}$$

and in the turbulent regime

$$f_{sl} = \left(\frac{0.079^8}{D^{*3}}\right)^{1/7} \tag{5.101}$$

The use of this plot is demonstrated in the following illustrative example.

Illustrative example 5.7

Calculate the rate of discharge of solid in the form of a slurry containing 9.5 per cent of solids by volume through a 5-cm diameter pipe under a pressure gradient of 91.8 Pa/m. The particle size in the slurry is 99 per cent $< 4\ \mu$m and independent measurements made in a rotational viscometer indicates that the fluid behaves as a Sisko fluid with flow index $n = 0.16$, $K_S = 0.7$ Pa s$^{0.16}$ and $\mu_\infty = 2.7 \times 10^{-3}$ Pa s. The density of the solid is 2970 kg/m^3 and the density of the water is 997 kg/m^3.

Solution

Since the pipe diameter is known the dimensionless pipe diameter is calculated

$$D^* = \left(\frac{1184 \times 91.8}{2 \times (2.7 \times 10^{-3})^2}\right)^{1/3} 0.05 = 97.67$$

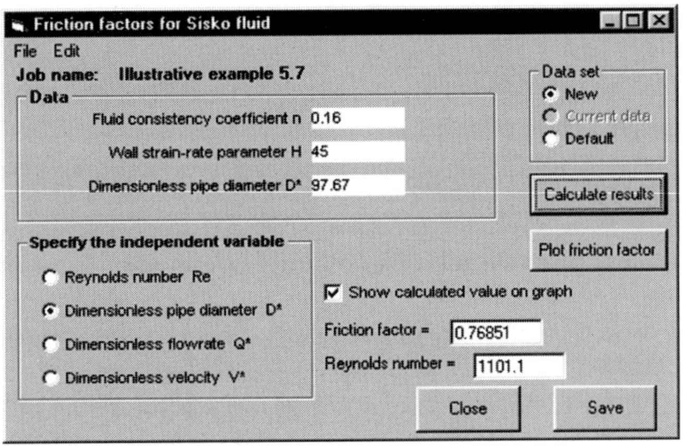

Figure 5.26 Toolbox data input screen to calculate friction factor for a Sisko fluid when the dimensionless pipe diameter is known

Even with a known value of D^*, trial and error methods cannot be avoided entirely for the Sisko fluid because the value of the strain-rate parameter H must be known in order to establish the value of the friction factor.

Try $H = 45$ to start the calculation and get the value of f_{sl} from a graph of the friction factor against D^* such as the one shown in Figure 5.25. Remember that a new graph must be generated for every different value of the flow index n. It is threfore essential to use the FLUIDS toolbox which also provides the value of Re_∞. The data input screen for the FLUIDS toolbox is shown in Figure 5.26.

$$f_{sl} = 0.7685$$
$$Re_\infty = 1101$$

The density of the slurry is calculated from the densities of the components

$$\rho_{sl} = 0.905 \times 997 + 0.095 \times 2970 = 1184\,\text{kg/m}^3$$

The average velocity is calculated from the Reynolds number

$$\bar{V} = \frac{1101 \times 2.7 \times 10^{-3}}{0.05 \times 1184} = 0.0502 \text{ m/s}$$
$$\frac{8\bar{V}}{D} = \frac{8 \times 0.0502}{0.05} = 8.034\,\text{s}^{-1}$$
$$H = \frac{0.7}{2.7 \times 10^{-3}} 8.034^{-0.84} = 45.1$$

which is close enough to the original guess.

The discharge rate of slurry is

$$Q = \frac{\pi}{4} D^2 \bar{V} = \frac{\pi}{4} \times 0.05^2 \times 0.0502 = 9.85 \times 10^{-5} \text{ m}^3\text{/s}$$

The discharge rate of solid is

$$Q_s = 0.095 \times 9.85 \times 10^{-5} = 9.36 \times 10^{-6} \ \text{m}^3/\text{s}$$

The mass discharge rate of solid is

$$W_s = \rho_s Q_s = 2970 \times 9.36 \times 10^{-6} = 0.0278 \text{ kg/s}$$

The dimensionless flowrate for a Sisko fluid is defined by

$$Q^{*3} = Re_{\infty}^{5} f_{sl} = \left(16 \frac{1+X}{G(n,X)}\right)^5 \frac{1}{f_{sl}^4} \tag{5.102}$$

and when the fluid is flowing steadily through a pipe, Q^* can be calculated from the flowrate and the available pressure gradient

$$Q^{*3} = \frac{32\,\rho_{sl}^4\,PGDTF}{\pi^3\,\mu_{\infty}} Q^3 \tag{5.103}$$

The dimensionless velocity for a Sisko fluid is defined by

$$V^{*3} = \frac{Re_{\infty}}{f_{sl}} = \frac{16}{f_{sl}^2} \frac{1+X}{G(n,X)} \tag{5.104}$$

and when the fluid is flowing steadily through a pipe, V^* can be calculated from the average velocity and the available pressure gradient

$$V^{*3} = \frac{2\,\rho_{sl}^2}{\mu_{\infty}\,PGDTF} \bar{V}^3$$

Graphs of the friction factor for Sisko fluids against Q^* and V^* can be easily generated using the FLUIDS toolbox.

5.5 Practice problems

1. Calculate the pressure gradient due to friction when a power law fluid flows through a 10-cm ID pipe at 20 kg/s.
   The properties of the fluid are:
   Density 1.7×10^3 kg/m^3
   K 1.2×10^{-3}
   n 0.23
2. The data given in the table below were obtained in a laboratory pipeline of diameter 0.053 m.

Pressure drop N/m^3	1000	14500	51000	100000
Q m^3/s	2.57×10^{-3}	1.25×10^{-2}	2.61×10^{-2}	3.89×10^{-2}

   Show that these data are consistent with the power-law model and evaluate the parameters n and K. The density of the fluid is 1000 kg/m^3.

3. Calculate the pipe diameter that is required to discharge 100 kg/hr of solid in the form of a slurry containing 9.5 per cent of solids by volume under a pressure gradient of 100 Pa/m. The particle size in the slurry is 99 per cent $< 4\mu$m and independent measurements made in a rotational viscometer indicates that the fluid behaves as a Sisko fluid with flow index $n = 0.16$, $K_S = 7.0$ Pa s$^{0.16}$ and $\mu_\infty = 2.7 \times 10^{-3}$ Pa s. The density of the solid is 2970 kg/m^3 and the density of the water is 997 kg/m^3.

5.6 Symbols used in this chapter

D	Pipe diameter m.
D^*	Dimensionless diameter.
f_{sl}	Friction factor for non-Newtonian fluid.
H	Strain-rate parameter for Sisko fluid.
He	Hedstrom number.
K	Flow consistency coefficient for power-law fluids Pa s^n.
K_H	Flow consistency coefficient for Herschel–Bulkley fluid pa s^n.
K_S	Flow consistency coefficient for Sisko fluid Pa s$^{n.}$
n	Flow behavior index for fluids having pseudo-plastic behavior.
$PGDTF$	Pressure gradient due to friction.
Q	Volumetric flowrate m^3/s.
Q^*	Dimensionless flowrate.
r	Radial position m.
Re	Reynolds number.
Re_B	Reynolds number for Bingham plastic.
Re_{PL}	Power-law Reynolds number.
Re_∞	Reynolds number for Sisko fluid.
u	Fluid velocity m/s.
$\bar{V}$	Average velocity m/s.
V^*	dimensionless velocity.
γ	Rate of strain s^{-1}.
γ_w	Rate of strain at the wall s^{-1}.
μ	Viscosity Pa s.
μ_B	Coefficient of plastic viscosity for a Bingham plastic Pa s.
μ_{eff}	Effective viscosity of a fluid defined as the ratio of local shear stress to local rate of strain Pa s.
μ_0	Lower limit of effective viscosity Pa s.
μ_∞	Viscosity at high strain rates Pa s.
ξ	Dimensionless variable.
ρ	Fluid density kg/m^3.
ρ_{sl}	Density of non-Newtonian fluid kg/m^3.

Bibliography

A comprehensive account of non-Newtonian fluids is given by Bird and Wiest (1996). The application of non-Newtonian fluid models to a range of industrial problems is discussed by Chabbra and Richardson (1999). Their text includes many interesting practical exercises that illustrate the application of the techniques that are discussed here. They also discuss the flow of non-Newtonian fluids in channels of non-circular cross-section. The measurement of rheological properties is discussed by Whorlow (1992).

The discussion of Sisko fluids is based on Turian *et al.* (1998a, b) who also present data on the friction loss when Sisko fluids flow through a variety of pipe fittings.

Chabbra, R.P. (1993) discusses the motion of bubbles, drops and particles in non-Newtonian fluids.

References

Bhattacharya, I.N., Panda, D. and Bandopadhyay, P. (1998). Rheological behavior of nickel laterite suspensions. *International Journal of Mineral Processing* **53**, 251–263.

Bird, R.B. and Wiest, J.M. (1996). Non-Newtonian liquids, Chapter 3 in *Handbook of Fluid Dynamics and Fluid Machinery*, Schetz, J.A. and Fuchs, A.E. (editors), pp. 223–302. John Wiley & Sons, New York.

Chabbra, R.P. (1993). *Bubbles, Drops and Particles in Non-Newtonian Fluids*. CRC Press.

Chabbra, R.P. and Richardson, J.F. (1999). *Non-Newtonian Flow in the Process Industries*. Butterworth-Heinemann.

Darby, R. (1988). Laminar and turbulent pipe flows of non-Newtonian fluids. Encyclopedia of Fluid Mechanics. Vol.7 Chapter 2, Cheremisinoff, N.P. (editor), Gulf Publishing Company.

Dodge, D.W. and Metzner, A.B. (1959). Turbulent flow of non-Newtonian systems. *A. I. Ch. E. Journal* **5**, 189–204.

Heywood, N.I. and Richardson, J.F. (1978). Rheological behavior of flocculated and dispersed aqueous Kaolin Suspensions in pipe flow. *Journal of Rheology* **22**, 599–613.

Huynh, L., Jenkins, P. and Ralston, J. (2000). Modification of the rheological properties of concentrated slurries by control of mineral-solution interfacial chemistry. *International Journal of Mineral Processing* **59**, 305–326.

Kemblowski, A. and Kolodziejski, J. (1973). Flow resistances of non-Newtonian fluids in transitional and turbulent flow. *International Chemical Engineering* **13**, 265–279.

Ma, T.-W. (1987). Stability, rheology and flow in pipes, bends, fittings, valves and venturi meters of concentrated non-Newtonian suspensions. PhD thesis, University of Illinois at Chicago.

Metzner, A.B. and Reed, J.C. (1955). Flow of Non-Newtonian fluids – correlation of the laminar, transition and turbulent-flow regions. *A. I. Ch. E. Journal* **1**, 434–440.

Thomas, D.G. (1960) Heat and momentum transport characteristics of non-Newtonian aqueous thorium oxide suspensions. *A I Ch E Journal* **6**, 631–639.

Turian, R.M., Ma, T.-W., Hsu, F.-L.G. and Sung, D.-J. (1998a). Flow of non-Newtonian Slurries: 1. Friction losses in laminar, turbulent and transition flow through straight pipes. *International Journal of Multiphase Flow* **24**, 243–269.

Turian, R.M., Ma, T.-W., Hsu, F.-L.G., Sung, D.-J. and Plackmann, G.W. (1998b). Flow of non-Newtonian Slurries: 1. Friction losses in bends, fittings, valves and venturi meters. *International Journal of Multiphase Flow* **24**, 225–242.

Whorlow, R.W. (1992). *Rheological Techniques*, 2nd edition, Ellis Horwood.

Wilhelm, R.H., Wroughton, D. M. and Loeffel, W.F. (1939) Flow of suspensions through pipes. *Industrial and Engineering Chemistry* **31**, 622–629.

6 Sedimentation and thickening

The ratio of water to solids in a slurry usually has a significant impact on the techniques and economics of the transport operations. Dilute slurries tend to behave in a fashion that is closer to that of Newtonian fluids while concentrated slurries can exhibit strong non-Newtonian behavior which the consequent effect on the energy that is required to pump the material at the required rate. Generally speaking there are usually advantages in reducing the amount of carrier fluid relative to the amount of solids to improve the energy efficiency as measured by the energy required to transport 1 kg of solids. Thus dewatering of slurries must always be considered in practice.

The solids content of a slurry can be increased using sedimentation techniques. The natural tendency of the solids to settle under the influence of gravity is exploited to remove some of the water from the slurry. When the particles that make up the slurry are small, the settling is quite slow and special techniques are required to achieve a separation. Because of the slow rates of settling that are commonly encountered, comparatively large equipment is required. In this chapter the basic principles of batch and continuous settling are discussed and applied to the analysis of the operation of industrial thickeners.

6.1 Thickening

Thickening is an important process for the partial dewatering of comparatively dense slurries. Essentially the slurry is allowed to settle under gravity but the particles are so close together that they hinder each other during settling and they tend to settle as a mass rather than individually. The rate of settling is a fundamental characteristic of the slurry which must be determined experimentally for each slurry under appropriate conditions of flocculation in order to design and size an appropriate thickener. The rate of settling depends on the nature of the particles that make up the slurry and on the degree of flocculation that is achieved. For a particular flocculated slurry the rate of settling is determined chiefly by the local solid content of the slurry and will vary from point to point in the slurry as the local solid content varies.

6.1.1 Batch thickening

The nature of the thickening process can be observed and the rate of settling determined as a function of the local solid content in a simple batch settling test. In this test the position of the liquid-solid interface, called the mudline, is observed as a function of time and the settling rate is determined as a function

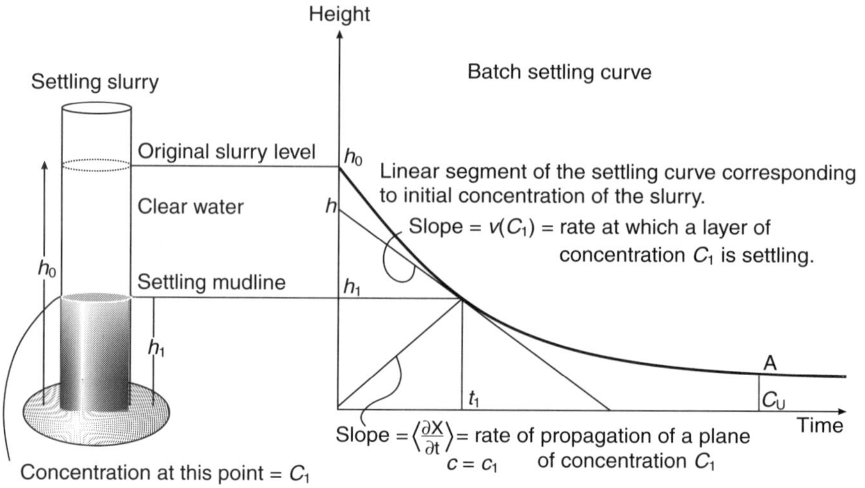

Figure 6.1 The batch settling test. A uniform slurry is allowed to settle under gravity and the level of the mudline is plotted as a function of time

of solid content using the so-called Kynch construction which is described below.

The batch settling test is illustrated in Figure 6.1. A sample of the flocculated slurry is gently and uniformly dispersed in the test cylinder and then allowed to settle undisturbed under gravity. If the flocculation is good, the slurry soon develops a well defined interface with clear water above and all the solids below. This interface falls as the solid particles below it settle under gravity. The sharpness of the interface is maintained because all of the particles at the interface are settling under hindered settling conditions and therefore settle at the same rate.

During the test, the position of the interface is measured and plotted as a function of time as shown in Figure 6.1. This simple test provides most of the information that is required to analyze, simulate and design a continuously operating thickener. After a long time the settling appears to stop and the interface falls very slowly while the sediment is compressing under its own weight. Eventually the settling stops altogether because the sediment reaches its ultimate concentration at which it is essentially *incompressible*. The concentration at which settling stops and the bed becomes incompressible is called the ultimate concentration and it is indicate as C_U in Figure 6.1. The position of the interface does not change after point A has been reached. The ultimate concentration corresponds to the situation when the solid particle that make up the settling sediment are closely packed in an arrangement that would correspond to hexagonal close packed if the particles were spherical in shape.

The method used to extract the information from the simple batch settling curve is known as the Kynch construction. The Kynch construction is applied to the batch settling curve and this establishes the relationship between the rate of settling of a slurry and the local solid content. The details of the Kynch

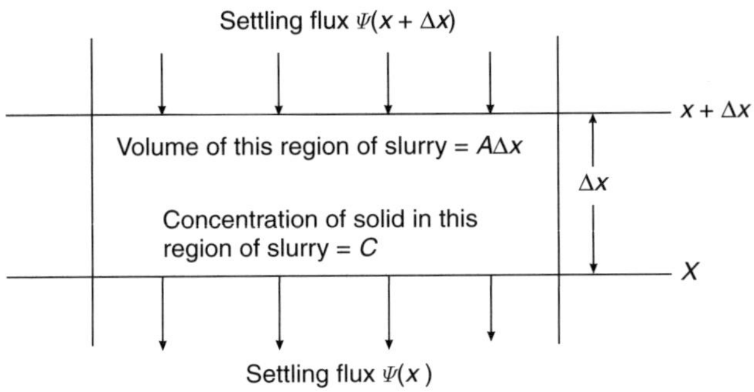

Figure 6.2 A typical slice through the settling slurry. The concentration of solids varies with time and position during the batch settling test

construction and its theoretical derivation from the basic Kynch hypothesis is given in detail below.

The settling flux, ψ, is the mass of solid crossing a stationary horizontal plane per unit time and specified per unit area of the plane. A differential mass balance within the settling slurry can be set up in terms of the flux, ψ, and the concentration, C, of solid at any horizontal plane in the settling portion of the slurry. A typical horizontal slice through the settling slurry is shown in Figure 6.2. The differential mass balance within the settling slurry relates the rate of change of concentration in the slice to the difference between the flux of solid entering from above and that leaving from below.

$$A(\psi(x+\Delta x)-\psi(x)) = A\Delta x\left(\frac{\partial C}{\partial t}\right)_x \tag{6.1}$$

In the limit as $\Delta x \to 0$ this gives the partial differential equation

$$\left(\frac{\partial \psi}{\partial x}\right)_t = \left(\frac{\partial C}{\partial t}\right)_x \tag{6.2}$$

In equations 6.1 and 6.2, the flux is assumed to be positive in the downward direction but the spatial coordinate, x, is taken to be positive in the upward direction.

During the batch settling test the concentration of solids is a function of both position and time and these variables are related by the following general expression

$$\left(\frac{\partial C}{\partial x}\right)_t \left(\frac{\partial x}{\partial t}\right)_C \left(\frac{\partial t}{\partial C}\right)_x = -1 \tag{6.3}$$

from which using equation 6.2

$$\left(\frac{\partial x}{\partial t}\right)_C = -\frac{\left(\frac{\partial C}{\partial t}\right)_x}{\left(\frac{\partial C}{\partial x}\right)_t} = -\frac{\left(\frac{\partial \psi}{\partial x}\right)_t}{\left(\frac{\partial C}{\partial x}\right)_t} = -\left(\frac{\partial \psi}{\partial C}\right)_t \tag{6.4}$$

Kynch postulated the fundamental principle that the settling velocity of the flocculated slurry is a function only of the local concentration of solids in the slurry. The settling velocity is the velocity relative to fixed laboratory coordinates that will be recorded if a single settling particle is observed. In terms of the Kynch postulate the settling velocity is not a function of particle size (true hindered settling) and not an explicit function of time or history of the settling slurry. The physical meaning of this postulate can be appreciated by considering some specific examples. During a batch settling test of the type that is illustrated in Figure 6.1 the concentration of solid in the slurry varies both with the level in the vessel and with the time. Consider that a concentration C_1 is achieved at some level at some particular time. If the slurry achieves the same concentration C_1 at some other level at another time during the test, the settling velocity of the solid at this concentration C_1 will be identical in both instances. It is also possible to conceive of doing two different experiments in two different settling vessels starting at two different initial concentrations and observing the settling velocity at any concentration C_1 that occurs in both tests at one or another time. The Kynch postulate asserts that the measured settling velocities would be identical. Clearly this implies a good deal of physical uniformity about the suspensions and in particular about the degree, structure and stability of the flocs that are usually essential for effective sedimentation of industrial suspensions. In spite of these very real limitations it is surprising how well the Kynch postulate can be satisfied in practice with careful preparation and testing of the slurry. A typical relationship between settling velocity and concentration is shown in Figure 6.3.

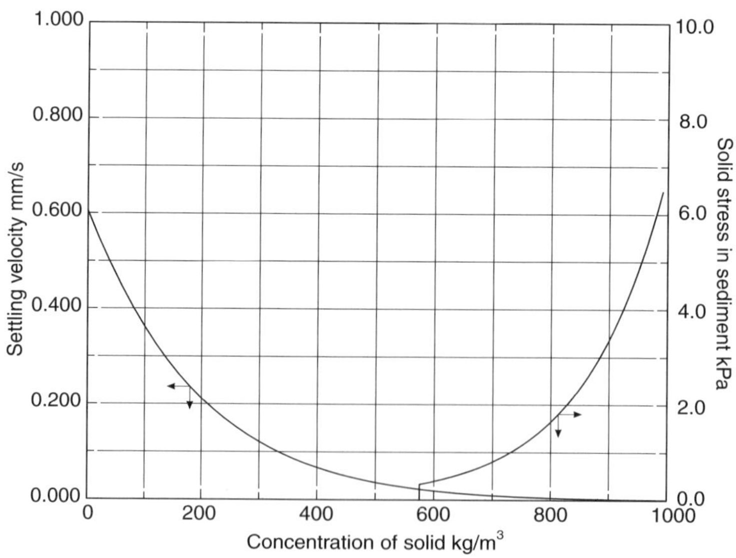

Figure 6.3 A typical representation of the relationship between the settling velocity and the solid concentration in the slurry as determined from the batch settling curve. The solid stress in a typical compressible sediment is also shown

The flux is related to the settling velocity by

$$\psi = CV(C) \tag{6.5}$$

$$\left(\frac{\partial \psi}{\partial C}\right)_t = C\left(\frac{\partial V(C)}{\partial C}\right)_t + V(C) = \text{function of } C \text{ only} \tag{6.6}$$

Thus we write

$$\left(\frac{\partial \psi}{\partial C}\right)_t = \frac{\mathrm{d}\psi}{\mathrm{d}C} = \psi'(C) \tag{6.7}$$

Using 6.4

$$\left(\frac{\partial x}{\partial t}\right)_C = -\left(\frac{\partial \psi}{\partial C}\right)_t = \text{function of } C \text{ only} \tag{6.8}$$

Thus $(\partial x/\partial t)_C$ is a function of C alone and therefore remains constant for each concentration C throughout the batch settling experiment. The derivative $(\partial x/\partial t)_C$ is the rate of propagation of a plane of constant concentration through the settling slurry and is related to the flux by equations 6.7 and 6.8

$$\left(\frac{\partial x}{\partial t}\right)_C = -\psi'(C) \tag{6.9}$$

The propagation velocity is positive (i.e. upward) wherever the function $\psi(C)$ has a negative gradient. During the batch settling test, the plane of constant composition C starts propagating from the bottom of the settling chamber at time zero. Since all concentrations up to the maximum are established there very soon after the start of the settling test, each concentration starts to propagate upward with the more concentrated planes usually moving slower than those less concentrated.

Now consider a particular plane having concentration C_1 moving upward through the slurry. The total transport of material through this plane gives the amount of solid in the slurry below the plane. When this plane hits the falling mudline, all the solid material in the original sample of slurry has passed through the plane.

This plane is moving upwards at a velocity $(\partial x/\partial t)_{C_1}$ while the solid is settling downwards through the plane at a velocity $V(C_1)$. Thus the total flux through this plane is then given by

$$\begin{aligned}\text{Flux} &= C_1\left(V(C_1) + \left(\frac{\partial x}{\partial t}\right)_{C_1}\right) \\ &= \text{constant}\end{aligned} \tag{6.10}$$

Total amount of solids that pass through this plane from the start of the experiment to time t_1 is given by

$$AC_1\left(V(C_1) + \left(\frac{\partial x}{\partial t}\right)_{C_1}\right)t_1 \tag{6.11}$$

and this must be equal to the total mass of solids in the slurry at time zero, AC_0h_0.

Observations of the falling mudline during the batch settling experiment provide values for the settling velocity but only at the mudline. This velocity can be obtained from the slope of the graph of interface position against time as shown in Figure 6.1.

Consider the time t_1 at which the plane of concentration C_1 arrives at the interface. This plane moves upwards at a constant rate $(\partial x/\partial t)_C$ as shown by a straight line from the origin in Figure 6.1.

The settling velocity at the interface is given by the slope of the settling curve there so that

$$V(C_1) = \frac{h - h_1}{t_1} \tag{6.12}$$

From equation 6.11

$$\begin{aligned} C_0\,h_0 &= C_1\left(V(C_1) + \left(\frac{\partial x}{\partial t}\right)_{C_1}\right)t_1 \\ &= C_1\left(\frac{h - h_1}{t_1} + \frac{h_1}{t_1}\right)t_1 \\ &= C_1\,h \end{aligned} \tag{6.13}$$

Equation 6.13 permits the calculation of the concentration at the interface at time t_1.

$$C_1 = \frac{C_0\,h_0}{h} \tag{6.14}$$

and this can be coupled with equation 6.12 to give a pair of points relating $V(C_1)$ and C_1. The analysis can be repeated at a number of different times and so build up a complete picture of the $V(C)$ vs C relationship and therefore the relationship between ψ and C. A typical graphical representation of this relationship is given in Figure 6.4.

In spite of the fact that the Kynch construction operates at the settling interface only, relationship between the settling flux and the concentration as shown in Figure 6.4 applies at every point in the slurry. C_C represents the concentration of an ideal fully settled incompressible sediment. When the concentration of the slurry reaches C_C, the flocs are in contact with each other and the entire floc bed is supported from below by the bottom of the test cylinder. If the sediment is incompressible no further settling takes place. In practice sediments are rarely incompressible because they continue to settle over time as water is expressed due to the compression of the flocs under the weight of the settled solids higher in the bed. It is not possible to determine C_C precisely in real slurries but nevertheless it is convenient to assume a single well-defined value for C_C in the simple theory. The ideal Kynch settling behavior occurs only in slurries having concentration less than the critical value C_C. Ideal Kynch settling is

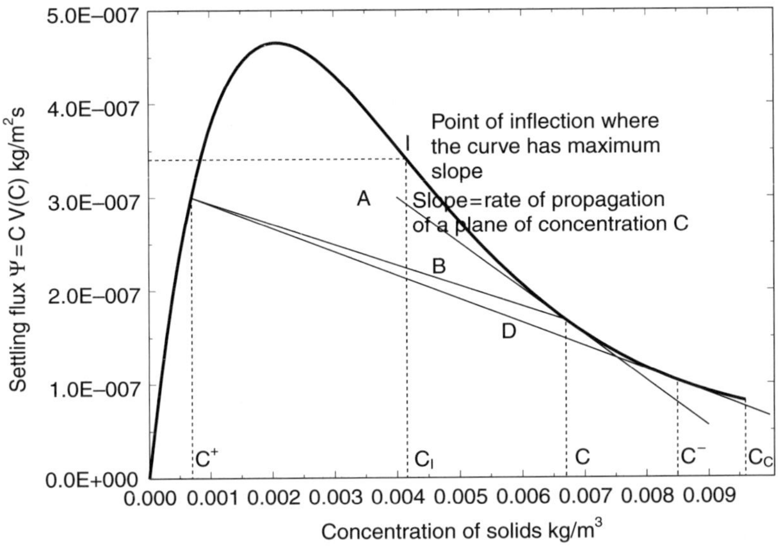

Figure 6.4 A typical relationship between the settling flux ψ and the concentration C

sometimes referred to as free settling behavior. The behavior of compressible sediments is discussed later in this chapter.

The point I on Figure 6.4 at which the flux curve has a point of inflection is particularly important. Equation 6.4 shows that the rate of upward propagation of a plane of constant composition is given by the negative partial derivative of $\psi(C)$ with respect to the concentration C. This can be obtained directly as the slope of the curve in Figure 6.4. The curve has its maximum slope at the point of inflection I so that a plane of concentration C_I will propagate faster than any other in the settler. If the batch settling test starts with a slurry having concentration less than C_I, the plane of concentration C_I moves upward from the floor of the settler and will overtake slower-moving planes ahead of it so that a discontinuity in the concentration develops quickly. This discontinuity moves upwards as a shock wave consuming solid above it until it strikes the downward-moving water-slurry interface and is destroyed. The plot of the interface position vs time will show a distinct corner at the point where the discontinuity meets the falling interface because of the abrupt change in the surface concentration which leads to an equally abrupted change in the settling velocity. This is shown in Figure 6.5. The shock wave is not often observed in practice because the test is usually performed only with slurries having an initial concentration larger than C_I. (True hindered settling is often not possible to achieve at lower concentrations). If the initial concentration C_0 is greater than C_I, the rates of propagation of all planes that originate at the bottom are lower than the plane having the initial concentration and no discontinuity develops. In this case the batch settling curve shows a smooth transition from the initial linear portion to the curved portion which develops after the arrival of the first plane that

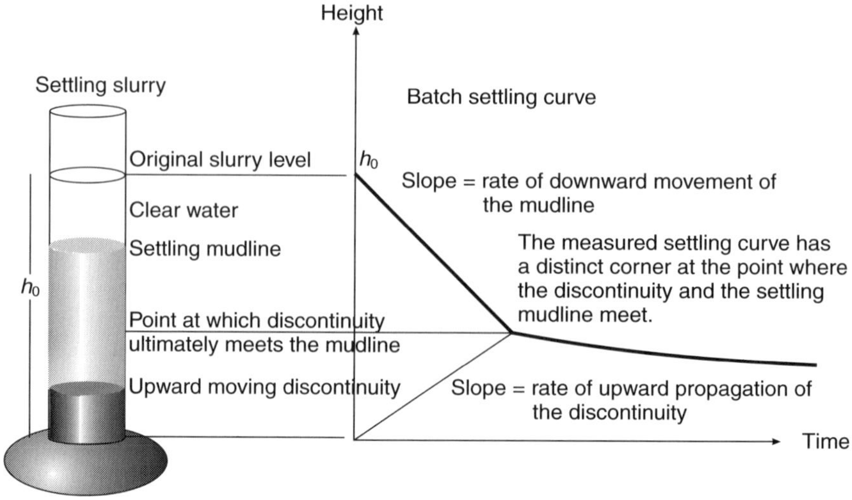

Figure 6.5 The batch settling test when the initial concentration of solids in the slurry is less than C_I. The batch settling curve has a distinct corner at the end of the constant rate period

orignates at the bottom of the settler at time zero and which has a concentration equal to the initial concentration C_0. This is illustrated in Figure 6.1.

It is possible to analyze the discontinuous case precisely and also the situation when the slurry has a flux-concentration relationship that shows more than one point of inflexion. A complete and useful mathematical analysis of this simple settling model is given by F. Concha and M.C. Bustos (1991).

6.2 Concentration discontinuities in settling slurries

The concentration discontinuities that develop in free settling slurries play an important role in describing the behavior of ideal slurries in a thickener. In the batch thickener a discontinuity is not stationary but moves up or down depending on the concentration just above and just below the discontinuity.

Consider a discontinuity as shown in Figure 6.6. The flux of solid into discontinuity from above

$$\psi(C^+) = C^+ V(C^+) \tag{6.15}$$

Flux of solid leaving from below

$$\psi(C^-) = C^- V(C^-) \tag{6.16}$$

In general

$$\psi(C^+) \neq \psi(C^-) \tag{6.17}$$

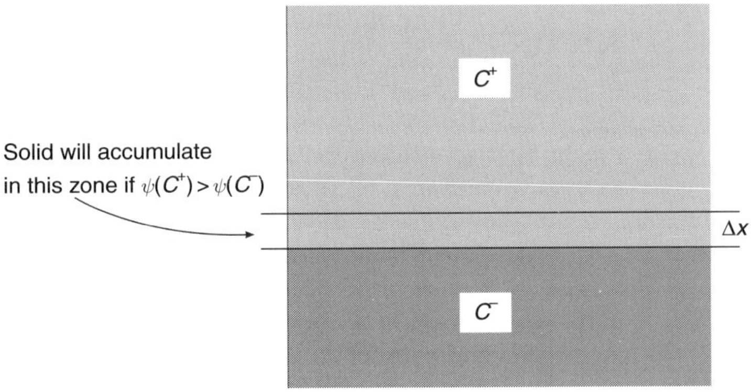

Figure 6.6 Discontinuities in the solid concentration can move upward or downward in a settling slurry

so that solid accumulates at the discontinuity. The accumulation is positive if

$$\psi(C^+) > \psi(C^-) \tag{6.18}$$

and vice versa. If the accumulation is positive, the discontinuity moves upward a distance Δx during a time interval Δt and these are related by

$$A\Delta x(C^- - C^+) = [\psi(C^+) - \psi(C^-)]A\Delta t \tag{6.19}$$

The rate of movement of the discontinuity is given by

$$\sigma(C^+, C\ ) = \lim_{\Delta t \to 0} \frac{\Delta x}{\Delta t} = -\frac{\psi(C^+) - \psi(C^-)}{C^+ - C^+} \tag{6.20}$$

A specific example of this condition is the rate at which the upper mudline drops at the start of the batch settling experiment. The concentration above the mudline is zero and the concentration immediately below the mudline is C_0, the initial concentration in the test. The velocity at which the mudline moves is given by equation 6.20

$$\sigma(0, C_0) = -\frac{\psi(0) - \psi(C_0)}{0 - C_0} = -\frac{\psi(C_0)}{C_0} = \frac{C_0 V(C_0)}{C_0} = V(C_0) \tag{6.21}$$

Thus the mudline discontinuity moves downward at the local solids settling velocity – an obvious fact that was used in equation 6.12.

An additional condition must be satisfied to ensure that a discontinuity can exist in practice. It must be hydrodynamically stable. To ensure this the density of the slurry above the discontinuity must be less than the density below the discontinuity. If this condition is not satisfied, convection currents would quickly develop at the interface and the sharp discontinuity would be destroyed. This requirement in turn means that the concentration of the slurry above the discontinuity must be lower than the concentration below the discontinuity because the density of a settling slurry always increases with increased concentration. Thus the condition

$$C^+ < C^- \tag{6.22}$$

must be satisfied at a stable discontinuity.

Inequality 6.22 is not sufficient to guarantee that a sharp discontinuity will actually form and persist somewhere in the settling slurry. This can be visualized from the following simple physical argument. A discontinuity in the concentration between regions with $C = C^+$ and $C = C^-$ can be considered to include all concentrations between C^- and C^+. If any of these concentrations, say C^*, can propagate faster than the discontinuity, the concentration immediately in front of the discontinuity will change to C^* and the sharp discontinuity between C^- and C^+ will be destroyed. A condition that ensures that this will not happen is

$$-\frac{\psi(C^*) - \psi(C^-)}{C^* - C^-} \leq -\frac{\psi(C^+) - \psi(C^-)}{C^+ - C^-} \tag{6.23}$$

for all values of C^* between C^- and C^+. In the mathematical literature this is known as an Oleinik condition.

This condition can also be written in another way by noting that it must apply to concentrations C^* that are arbitrary close to C^- and C^+ but within the interval (C^+, C^-). Consider the concentration $C^* = C^- - \Delta C$ with $\Delta C > 0$. Condition 6.23 requires

$$\begin{aligned}
\sigma(C^+, C^-) &\geq -\frac{\psi(C^- - \Delta C) - \psi(C^-)}{-\Delta C} \\
&\geq \frac{\psi(C^-) - \psi'(C^-)\Delta C + o(\Delta C) - \psi(C^-)}{\Delta C} \\
&\geq -\psi'(C^-) + \frac{o(\Delta C)}{\Delta C}
\end{aligned} \tag{6.24}$$

which implies that

$$\psi'(C^-) \geq -\sigma(C^+, C^-) \tag{6.25}$$

at a stable discontinuity.

A similar analysis at C^+ gives

$$\psi'(C^+) \geq \sigma(C^+, C^-) \tag{6.26}$$

These requirements are illustrated in Figure 6.4. Consider a sedimenting slurry having a region with concentration C^+. This region can be in contact with a slurry of greater concentration across a discontinuity. This concentration, called the conjugate concentration, will be at some concentration on the falling section of the flux curve. However, the conjugate concentration must satisfy all of the conditions 6.23, 6.25 and 6.26. The slope of the tie line B that connects the two conjugate concentrations on the flux curve is the negative of the rate of propagation of the discontinuity. For example, if the concentration conjugate to C^+ is at C, the condition 6.25 will not be satisfied and a stable discontinuity will not develop between these two concentrations. The only concentration that can be conjugate across a stable discontinuity is C^- where the conjugate tie line D is tangential to the flux curve so that condition 6.25 is satisfied as an equality.

6.3 Useful models for the sedimentation velocity

A number of models have been proposed to relate the sedimentation velocity to the solids concentration and these have been used with varying success to describe the sedimentation of a variety of slurries that are important in industrial practice. It is usual to express these model equations in terms of the volumetric concentration of solid φ expressed as m^3 solid/m^3 slurry. The two representations of the solid concentration are related by

$$C = \rho_s \varphi \tag{6.27}$$

where ρ_s is the density of the solid.

The most commonly used model is based on the work of Richardson and Zaki (1954)

$$V = \nu_{TF}(1 - r_F \varphi)^n \quad \text{for } \varphi < \frac{1}{r_F} \tag{6.28}$$

Here ν_{TF} is the terminal settling velocity of an isolated floc and r_F is the floc dilution which is the volume of a floc per unit volume of contained solids. The recommended value for n is 4.65. This equation has been applied to the sedimentation behavior of flocculated Witwatersrand gold ore slurries and some typical data are compared to this model in Figure 6.7 with $\nu_{TF} = 0.778$ mm/s and $r_F = 0.973\,\text{m}^3$ floc/t solid. The following values have been reported for Kaolin slurries at pH 6.0: $\nu_{TF} = 0.36$ mm/s and $r_F = 40\,\text{m}^3$ floc/m^3 solid. Shirato *et al.* (1970) reported the following values for zinc oxide slurries: $\nu_{TF} = 0.715$ mm/s and $r_F = 200\,\text{m}^3$ floc/m^3 solid and for ferric oxide slurries $\nu_{TF} = 6.42$ mm/s and $r_F = 17\,\text{m}^3$ floc/m^3 solid.

An alternative model has also been used by Shirato *et al.*

$$V = \alpha \frac{(1-\varphi)^3}{\varphi} \exp(-\beta\varphi) \tag{6.29}$$

Zinc oxide slurries in the range $0.02 \le \varphi \le 0.26$ have been described using $\alpha = 5.7 \times 10^{-5}$ mm/s and $\beta = 6.59$.

Careful experimental work over wide concentration ranges by Wilhelm and Naide (1981) has shown that the settling velocity of flocculated slurries over restricted concentration ranges follows the power law

$$\frac{1}{V} = a\,C^b \tag{6.30}$$

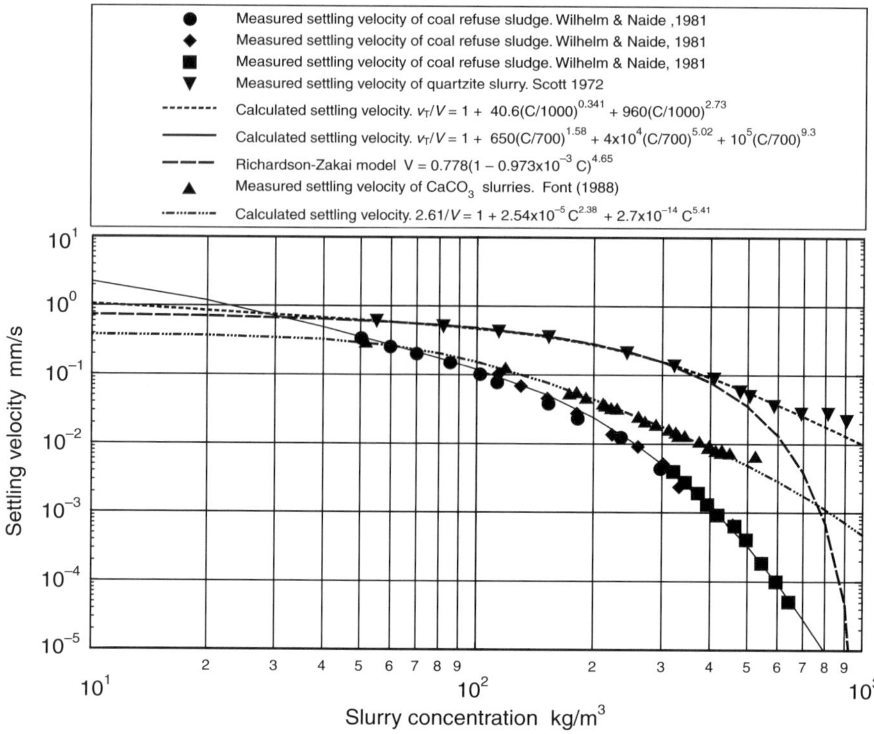

Figure 6.7 Experimental data for flocculated slurries obtained using the Kynch graphical construction and fitted by the extended Wilhelm–Naide equation

Therefore the complete settling-velocity concentration relationship follows a simple equation

$$\frac{\nu_{TF}}{V} = 1 + \alpha_1 C^{\beta_1} + \alpha_2 C^{\beta_2} + \cdots \tag{6.31}$$

with $0 < \beta_1 < \beta_2 < \cdots$. Each term dominates the expression over a limited concentration range. This equation describes the settling behavior of a variety of slurries over wide ranges of C. Values for β_i range from 1.0 to as high as 20 or more. Usually no more than two power-function terms are required in this equation to describe the settling velocity over two or more orders of magnitude variation in the concentration. In equation 6.31, ν_{TF} represents the terminal settling velocity of the individual flocs when they are widely separated from each other and settle as single entities. ν_{TF} is quite difficult to measure but its value can be estimated using the methods of Chapter 3 if the size and effective density of the individual flocs can be measured. Equation 6.31 will be referred to as the extended Wilhelm-Naide equation.

Experimental data on the settling of coal refuse sludge measured by Wilhelm and Naide and of flocculated quartzite slurries typical of mineral processing operations are shown in Figure 6.7 where they are compared against equation

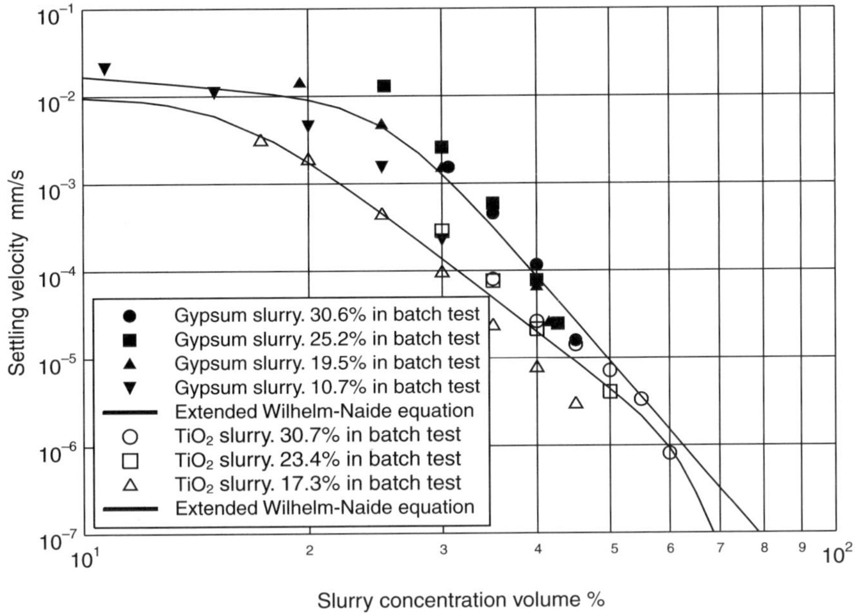

Figure 6.8 Measured settling velocities for unflocculated slurries. The lines are calculated using the Extended Wilhelm–Naide equation. Data is from Ma (1987)

6.31. Equation 6.31 describes the settling behavior of these flocculated slurries well and for these data is significantly better than the Richardson-Zaki model.

This model is also useful for unflocculated slurries and a comparison is shown in Figure 6.8 using batch settling data from Ma (1987) which is also presented in Turian *et al.* (1997). The original data were analyzed using the Kynch graphical construction and are shown as settling velocity as a function of the solids concentration. In spite of the inevitable scatter in the data that is associated with the batch settling test method and the subsequent graphical construction, the extended Wilhelm-Naide equation provides an appropriate representation of the data. The TiO_2 slurry was made from particles having $d_{80} = 1.89\,\mu\text{m}$, $d_{50} = 0.89\,\mu\text{m}$ and $d_{10} = 0.4\,\mu\text{m}$. The gypsum suspension was made from coarser particles having $d_{80} = 99.8\,\mu\text{m}$, $d_{50} = 38.3\,\mu\text{m}$ and $d_{10} = 7.29\,\mu\text{m}$. The parameters in the extended Wilhelm-Naide equation for these suspensions are given by

$$\frac{\nu_{\text{TF}}}{V} = 1 + 34.05\varphi^{0.855} + 1.1 \times 10^7 \varphi^{10} \tag{6.32}$$

and

$$\frac{\nu_{\text{TF}}}{V} = 1 + 2.38 \times 10^5\,\varphi^{6.73} + 7.26 \times 10^8\,\varphi^{24.2} \tag{6.33}$$

for the gypsum and TiO_2 respectively. The concentration is expressed as volume fraction in both cases.

In spite of its completely empirical nature, the extended Wilhelm-Naide equation provides a versatile and flexible description of the settling velocity. It is used in the FLUIDS toolbox and has proved to be the most useful equation to describe experimental thickening data.

6.4 Continuous cylindrical thickener

A cylindrical ideal thickener that operates at steady state is shown schematically in Figure 6.9. The feed slurry is introduced below the surface and a sharp interface, A, develops at the feed level between the clear supernatant fluid and a slurry of concentration: C_L. The feed slurry is assumed to spread instantly across the cross-section of the thickener and to dilute to concentration C_L.

Obviously this is an idealization of the behavior in a real thickener. Nevertheless it provides a useful simulation model. Lower down in the thickener, an interface, B, having concentration C_M develops and at the bottom of the thickener the mechanical action of the rake moves the settled pulp inward and the fully thickened slurry is discharged through the discharge pipe at concentration C_D. The solid concentration of the sediment below interface B is not uniform and it increases with depth due to the compressibility of the pulp.

The operation of the thickener is dominated by the behavior of these layers and the relationships between them. The concentration of solids in each of the layers is constrained by the condition that the thickener must operate at steady state over the long term. If the slurry is behaving as an ideal Kynch slurry, well-defined sharp interfaces will develop in the thickener and the analysis below shows how these concentrations can be calculated.

The total settling flux relative to fixed coordinates at any level where the concentration is C must include the effect of the net downward volumetric flow that is due to the removal of pulp at the bottom discharge in addition to the settling flux of the solid relative to the slurry itself. If the total flux is

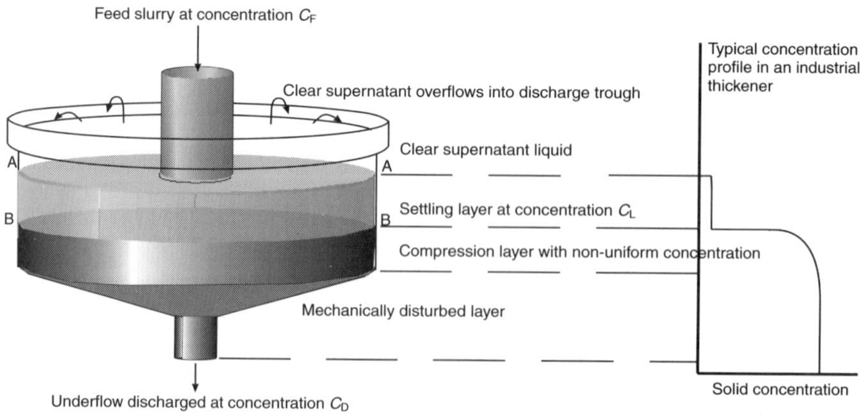

Figure 6.9 Schematic representation of the ideal thickener operating at steady state

represented by *f(C)* and the volumetric flux of slurry below the feed by q m^3/m^2 then

$$f(C) = qC + \psi(C)\,\text{kg/m}^2\text{s} \tag{6.34}$$

In batch settling $q = 0$ so that *f(C)* and $\psi(C)$ are identical. *f(C)* is plotted for different values of q in Figure 6.10 using the data of Figure 6.3.

The analysis leading to equation 6.20 for the batch settler can be used to develop an expression for the rate at which a discontinuity will move in a continuous thickener.

$$\sigma(C^+, C^-) = -\frac{f(C^+) - f(C^-)}{C^+ - C^-}\ \text{m/s} \tag{6.35}$$

where C^+ and C^- represent the concentrations of solids above and below the discontinuity respectively.

If the thickener is operating at steady state, the discontinuities must not move and $\sigma(C^+, C^-)$ must be zero across every discontinuity. The right hand side of equation 6.35 is the negative of the slope of the chord connecting two points on the flux curve and these chords must be horizontal to satisfy the steady state requirement. The concentrations in the layers on each side of a discontinuity make up a conjugate pair. These conjugate concentrations are further limited but the requirement that all concentration discontinuities

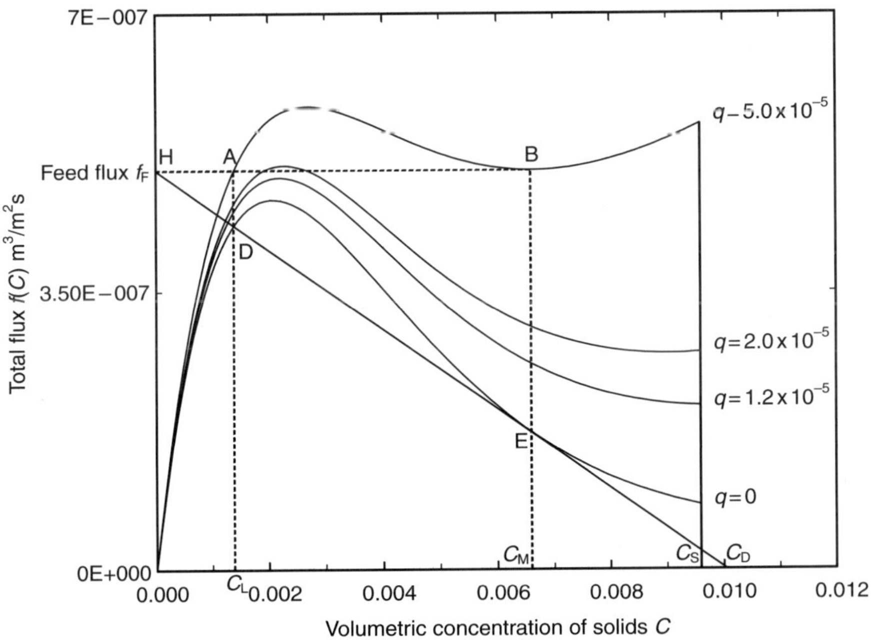

Figure 6.10 Graphical procedure to describe the steady state operation of an ideal thickener

must be stable as well as stationary. The stability of the interface requires that the higher conjugate concentration can exist only at concentration C_M at which the total settling flux has a local minimum so that condition 6.25 is satisfied with $f(C)$ replacing $\psi(C)$. Thus as soon as the underflow volumetric flux q is fixed the conjugate concentrations can be determined by drawing the horizontal tangent to the total flux curve as shown by line A–B in Figure 6.10.

The flux curves shown in Figure 6.10 can be used to develop a simple ideal model of the continuously operating cylindrical thickener. The model is based on the requirement that at the steady state all the solid must pass through every horizontal plane in the thickener. In other words, the solid must not get held up anywhere in the thickener. If that were to happen, solid will inevitably accumulate in the thickener which will eventually start to discharge solids in its overflow.

The flux through any horizontal plane in a steady state thickener must equal the feed flux

$$f_F = \frac{Q_F C_F}{A} \tag{6.36}$$

and the underflow flux

$$f_D = \frac{Q_D C_D}{A} = qC_D \tag{6.37}$$

where $q = Q_D/A$ is the total downward volumetric flux at any horizontal layer below the feed well.

Thus

$$f(C) = qC + \psi(C) = f_F \tag{6.38}$$

where C is the concentration at any level where free settling conditions exist in the thickener.

Equation 6.38 can be plotted on the $\psi(C)$ vs C axes as a straight line of slope $-q$ as shown as line HDE in Figure 6.10 which is plotted for the case $q = 5.0 \times 10^{-5}$ m/s.

Re-arranging equation 6.38

$$\psi(C) = f_F - qC \tag{6.39}$$

which shows that the line intersects the $\psi(C)$ axis at $\psi = f_F$ and since

$$f_D = qC_D = f_F \tag{6.40}$$

it intersects the C axis at $C = C_D$.

The quantity $qC + \psi(C)$ is sometimes referred to as the demand flux because this flux must be transmitted through every horizontal level in the thickener otherwise the thickener could not operate at steady state.

It is not difficult to show that the straight line representing equation 6.38 is tangential to the $\psi(C)$ function at point E which has the same abscissa value

C_M as the conjugate operating point B. The point E is directly below B in Figure 6.10.

At point B

$$\begin{aligned} \frac{df(C)}{dC} &= q + \frac{d\psi(C)}{dC} = 0 \\ \frac{d\psi(C)}{dC} &= -q \end{aligned} \tag{6.41}$$

Thus $C_B = C_E = C_M$.

Likewise the intersection D between the straight line and the $\psi(C)$ curve has the same concentration as the lower conjugate point A.

Point A is defined by

$$f_F = qC_A + \psi(C_A) \tag{6.42}$$

and the point D is defined by

$$f_F - qC_D = \psi(C_D) \tag{6.43}$$

These two equations are true simultaneously only if $C_D = C_A = C_L$.

A simple simulation model can be constructed for the continuous cylindrical thickener using this ideal model. If the area of the thickener is given and the conditions in the feed pulp are known then

$$f_F = Q_F C_F \tag{6.44}$$

Thus the maximum underflow concentration and the underflow pumping rate is fixed by the abscissa intercept and by the slope of the line HDE in Figure 6.10. This defines the flowrate and the composition of the pulp that is passed from the thickener underflow.

The maximum possible feed flux is fixed by the slope of the flux curve at the point of inflexion ψ'_I.

$$f_{F\max} = \psi(C_I) + C_I\psi'_I \tag{6.45}$$

The method requires that a suitable model be available for the settling flux. This can be obtained from the batch settling curve as described in section 6.2.1 or from a model of the settling velocity. The Richardson-Zaki model for the sedimentation velocity can be used to build a simple but self-consistent simulation model for the ideal thickener. The maximum feed rate of solid that can be sent to a thickener of given diameter is fixed by the slope of the sedimentation flux curve at the point of inflexion. The sedimentation flux is given by equation 6.5

$$\psi = \nu_{TF} C(1 - r'_F C)^n \tag{6.46}$$

where

$$r'_F = \frac{r_F}{\rho_s} \tag{6.47}$$

The point of inflexion is at

$$C_I = \frac{2}{r'_F(n+1)} \tag{6.48}$$

and the critical slope at the point of inflexion is given by

$$\psi'_I = \frac{\nu_{TF}}{(n+1)^n}\left((n-1)^n - 2n(n-1)^{n-1}\right) \tag{6.49}$$

The maximum possible feed flux occurs when the operating line in Figure 6.10 is tangential to the flux curve at the critical point of inflexion. Thus

$$\begin{aligned} f_{F\max} &= \psi(C_I) - \psi'_I C_I \\ &= \frac{4\nu_{TF}}{r'_F} \times \frac{n(n-1)^{n-1}}{(n+1)^{n+1}} \end{aligned} \tag{6.50}$$

The maximum possible feed rate of solids to the thickener is

$$W_{F\max} = \rho_s f_{F\max} A \text{ kg/s} \tag{6.51}$$

When the thickener is fed at a rate less than the maximum, the maximum concentration of the underflow can be calculated from the intersection with the horizontal axis of the operating line that passes through the given feed flux on the vertical axis and which is tangential to the flux curve as shown in Figure 6.10. This requires the solution of a non-linear equation

$$\begin{aligned} C_M &= \frac{\psi(C_M) - f_F}{\psi'(C_M)} \\ &= \frac{C_M(1 - r'_F C_M)^n - \dfrac{f_F}{\nu_{TF}}}{(1 - r'_F C_M)^n - n r_F C_M (1 - r'_F C_M)^{n-1}} \end{aligned} \tag{6.52}$$

for the intermediate concentration C_M after which the maximum concentration of the pulp in the discharge is calculated from

$$\frac{C_D}{C_M} = \frac{f_F}{f_F - \psi(C_M)} \tag{6.53}$$

The smallest volumetric discharge rate that is possible for steady-state operation is calculated from

$$q_M = \frac{f_F}{C_D} \tag{6.54}$$

This calculation is illustrated in Illustrative example 6.1.

Illustrative example 6.1

The free settling behavior of a slurry is governed by a Richardson-Zaki equation.

$V(\varphi) = 0.605(1 - \varphi)^{12.59}$ mm/s

The density of the solid is 2500 kg/m^3.

Calculate the maximum feed rate that can be handled by a 50-m diameter thickener.

Calculate the maximum discharge concentration if the thickener is fed at 100 kg/s.

Solution

The point of inflection on the flux curve is given by equation 6.48,

$$C_I = \frac{2}{r'_F(n+1)} = \frac{2 \times 2500}{12.59 + 1} = 367.9\,\text{kg/m}^3$$

The maximum feed flux is given by equation 6.50,

$$f_{F\max} = \frac{4 \times 0.605 \times 10^{-3} \times 2500 \times 12.59 \times 11.59^{11.59}}{13.59^{13.59}} = 0.0652\,\text{kg/m}^2\text{s}$$

The maximum feed rate that can be handled is

$$\begin{aligned} F_{\max} &= 0.0652 \times \frac{\pi}{4} \times 50^2 \\ &= 128.0\,\text{kg/s} \\ &= 460.9\,\text{t/h} \end{aligned}$$

The intermediate concentration C_M for feed rate 100 kg/s is calculated using equation 6.52.

The feed flux is

$$f_f = \frac{100}{\frac{\pi}{4} 50^2} = 0.0509\,\text{kg/m}^2\text{s}$$

Equation 6.52 is

$$C_M = \frac{C_M - \dfrac{0.0509}{0.605 \times 10^{-3}\left(1 - \dfrac{C_M}{2500}\right)^{12.59}}}{1 - \dfrac{12.59\,C_M}{2500\left(1 - \dfrac{C_M}{2500}\right)}}$$

This equation can be solved by iteration to give $C_M = 556.9\,\text{kg/m}^3$. The maximum discharge concentration is obtained from equation 6.53.

$$\psi(C_M) = 0.605 \times 10^{-3} \times C_M(1 - r'_F C_M)^{12.59} = 0.0141\,\text{kg/m}^2\text{s}$$

$$\begin{aligned} C_{D_{\max}} &= \frac{556.9 \times 0.0509}{0.0509 - 0.0141} \\ &= 770.3\,\text{kg/m}^3 \end{aligned}$$

The volumetric discharge rate is given by equation 6.54,

$$
\begin{aligned}
q &= \frac{0.0509}{770.3} \\
&= 6.608 \times 10^{-5}\,\text{m}^3/\text{m}^2\text{s} \\
Q &= 6.603 \times 10^{-5} \frac{\pi}{4} 50^2 \\
&= 0.1297\,\text{m}^3/\text{s}
\end{aligned}
$$

When the more useful and widely applicable extended Wilhelm-Naide model is used for the sedimentation velocity, the analytical method used above does not produce nice closed-form solutions, and numerical methods are required to solve the equations iteratively. The FLUIDS toolbox can be used to do these computations conveniently.

The construction illustrated in Figure 6.10 provides a rapid and simple design procedure for an ideal thickener based on the ideal theory. Either the maximum underflow concentration or the feed concentration can be specified and the other is fixed by the line drawn tangent to the settling flux curve. This also fixes the minimum total volumetric flux q from which the required area of the thickener can be determined.

$$A = \frac{Q_D}{q} = \frac{Q_F C_F}{C_D q} \tag{6.55}$$

Under the assumption that the settled pulp is incompressible, the maximum discharge pulp concentration is C_C and when the thickener discharges at this concentration, only one feed flux and one volumetric flux q is possible for the thickener as shown in Figure 6.10. In practice the thickener discharge concentration is always greater than the concentration at the lower conjugate concentration C_M although the simple Kynch theory provides no mechanism to describe how the concentration increases from C_M to C_D. In practice the sediment will always be compressible and natural compression of the sediment causes a steady increase in the solid concentration with sediment depth. Additional modeling considerations are required to describe the compression process. These are discussed in Section 6.6.

6.5 Simulation of the batch settling experiment

The batch settling experiment can be used to determine the sedimentation velocity–concentration relationship by means of the Kynch graphical construction that was described in Section 6.1. However, it is usually more convenient to simulate the batch settling experiment using the chosen sedimentation velocity model and to compare the simulation with the measured batch height vs time curve directly. The parameters in the sedimentation velocity model can then be estimated using standard parameter estimation techniques.

The height of the mudline in the batch settling experiment can be calculated simply by integrating the settling velocity at the concentration that is present at the mudline at any point during the experiment.

Thus referring to Figure 6.1

$$-\frac{dh}{dt} = V(C_1) \tag{6.56}$$

Where C_1 is the concentration at the mudline. Recall that the Kynch analysis given in Section 6.1 shows that C_1 is not constant but varies during the course of the batch settling experiment. The experiment consists of three distinct time periods. The first extends from the start of the experiment to the time t' at which the plane of concentration C_0 (the initial uniform starting concentration) hits the falling mudline. During the interval $t = 0$ to $t = t'$, the concentration at the mudline is equal to C_0 which remains fixed and the mudline falls at constant velocity. This is called the constant rate period and the graph of h vs t is a straight line.

The time t' at which the constant rate period ends can be calculated simply using the geometry of Figure 6.1. The initial constant rate period of the settling curve is a straight line from the starting point h_0, to the intersection with the line of propagation of a plane of concentration C_0 from the bottom of the settling cylinder. Thus the coordinates (t', h') can be calculated from the simultaneous solution of the two equations

$$\frac{h' - h_0}{t'} = -V(C_0) \tag{6.57}$$

which represents the falling mudline and

$$\frac{h'}{t'} = -\frac{d\psi(C_0)}{dC} = \psi'(C_0) \tag{6.58}$$

which represents the upward moving plane of concentration C_0.

The solution is

$$t' = -\frac{h_0}{V(C_0) - \psi'(C_0)} \tag{6.59}$$

If C_0 is lower than the concentration at the point of inflection in the settling curve, then

$$t' = -\frac{h_0}{V(C_0) + R'(C_0)} \tag{6.60}$$

where $R'(C_0)$ is the rate of propagation of a discontinuity into a slurry of concentration C_0. This rate is given by $\sigma(C_0, C^*)$ where C^* is the concentration immediately below the discontinuity. The geometry of the flux curve limits this to a single value, C^*, since the only tie line that connects the initial concentration C_0 to another point on the curve with $C > C_0$ must be tangential to satisfy condition 6.25.

The height of the interface at the end of the constant rate period is

$$h' = h_0 - V(C_0)t' \tag{6.61}$$

After the end of the constant rate period, the concentration of solid at the mudline increases steadily as the settling proceeds. At any time $t > t'$ the mudline is at coordinate (t, h) and the concentration is given by

$$C = G^{-1}\left(-\frac{h}{t}\right) \tag{6.62}$$

where G^{-1} is the inverse of the function $\psi'(C)$ and C is the concentration at height h at time t that results from the upward propagation of the plane from the floor of the settling cylinder.

The rate at which the mudline falls is given by

$$-\frac{dh}{dt} = V\left[G^{-1}\left(-\frac{h}{t}\right)\right] \tag{6.63}$$

Equation 6.63 can easily be solved numerically using any standard technique for the numerical solution of ordinary differential equations. The solution is started from the initial condition $h = h'$ at $t = t'$ and $C = C_0$. A simulated batch settling curve is compared with experimental data in Figure 6.11.

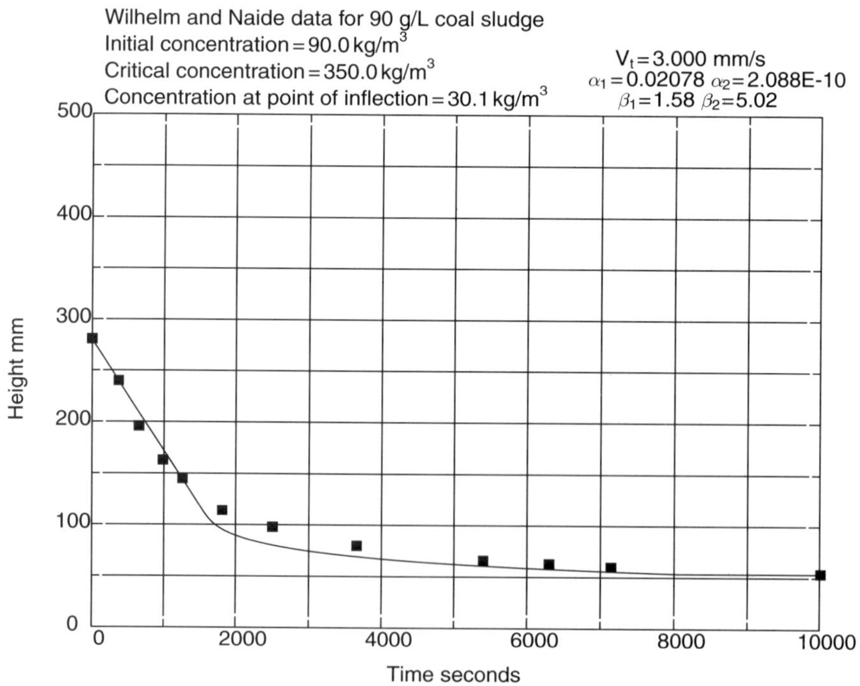

Figure 6.11 Simulated batch settling curve using the extended Wilhelm–Naide model for the settling velocity. Experimental data from Wilhelm and Naide 1981

In the event that $C_0 < C_I$ (the point of inflection on the flux curve as shown in Figure 6.4) the settling curve has a discontinuity in the slope at (t', h') as shown in Figure 6.5 In this case the numerical solution is started from (t', h') and $C = C_I$.

When the concentration at the mudline reaches the critical concentration C_C, sedimentation stops if the sediment is incompressible and

$$\frac{dh}{dt} = 0 \tag{6.64}$$

The final state of the batch test is a sediment of concentration C_C having a height

$$h_z = \frac{C_0 L}{C_z} \tag{6.65}$$

with a layer of clear water on top.

6.6 Thickening of compressible pulps

Figure 6.12 shows measured pulp density profiles in an industrial thickener operating normally and in an overloaded condition. The measured profile

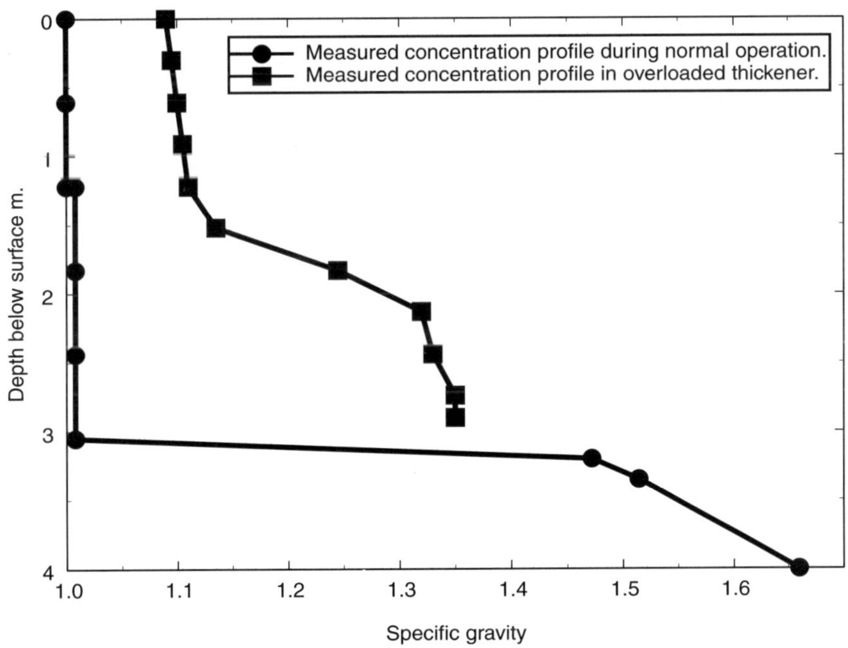

Figure 6.12 Measured density profiles in an industrial thickener. The specific gravity of the feed was 1.116 and the underflow discharged at a specific gravity of 1.660 under normal conditions. Data is from Cross (1963)

during normal operation agrees with that expected in an ideal thickener and the gradual increase in the pulp concentration between the lower conjugate concentrate C_M and the discharge concentration C_D is clearly evident. The data is from Cross (1963). The thickener was 22.9 m in diameter with a 3 m cylindrical section and a cone depth of 1.55 m.

Ideal Kynch thickening behavior does not describe the entire thickening process because it is not applicable when the sediment is under compression at the bottom of the thickener. Ideal Kynch thickening terminates when the slurry concentration is sufficiently high to allow individual flocs to touch and support each other. Consequently there is no natural settling of the individual particles relative to the water. However, as the floc bed increases in height, the weight of the accumulated flocs compresses the lower layers of the floc bed and the water is squeezed out. This water is forced upward through the floc bed and the upward drag on the flocs actually helps to support them. Even if the individual flocs are not themselves compressible, groups of flocs exhibit compressive behavior in that compression forces express water from the voids between the flocs. The touching flocs generate a structure that has internal strength which is a function of the solid concentration. This internal strength manifests itself as a normal stress on the solid phase and it is this stress that supports the upper layers of flocs in the compression zone (see Figure 6.13).

While the floc bed is being compressed, the interstitial water flows upward through the floc bed. The viscous drag generated by this upward flowing water helps to support the layers of flocs. The concentration at which the flocs just form supportive contacts is called the critical concentration C_C. This concentration plays a pivotal role in determining the behavior of both batch and continuous thickeners. In the upper part of the thickener where the slurry concentrations are less than C_C, ideal Kynch behavior occurs. In the lower regions of the thickener where slurry concentrations are greater than C_C, a different model is required to describe the behavior of the sediment. In particular the settling velocity of the flocs in the sediment depends not only on the local concentration of the solids but on the gradient of the concentration as well, a condition which violates the basic Kynch postulate. A force

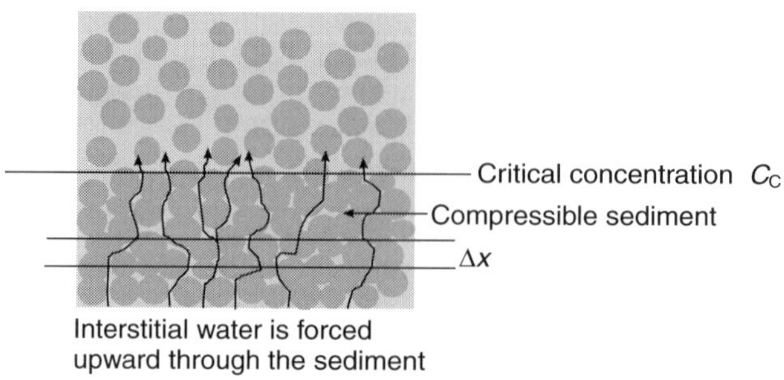

Figure 6.13 The interface between free settling slurry and the compressible sediment

balance over a horizontal slice of compressible floc bed will generate an equation for the rate of increase in the solid stress at increasing depth.

$$\sigma_{x+\Delta x}A + F_g A\Delta x = \sigma_x A + F_d A\Delta x \tag{6.66}$$

where F_g and F_d represent respectively the gravitational and drag forces on the solid phase per unit volume of sediment. In the limit this equation becomes

$$-\frac{\partial \sigma}{\partial x} = F_g - F_d \tag{6.67}$$

The gravitational and drag forces are modeled as follows

$$F_g = (\rho_s - \rho_f)g\varphi = \Delta\rho g\varphi \,\mathrm{N/m^3} \tag{6.68}$$

$$F_d = \frac{\mu V_r \varphi}{K}\,\mathrm{N/m^3} \tag{6.69}$$

Where K is the permeability of the flocculated sediment, μ the viscosity of the fluid and V_r the interstitial velocity of the water relative to the solid. The drag force can be expressed in terms of the terminal settling velocity of the floc bed by noting that if a slice of the bed were free to settle under its own weight (no support from below or pressure from above and no friction on the vertical walls) it would do so at its terminal settling velocity. Under these conditions

$$\frac{\mu V_T \varphi}{K(1-\varphi)} = F_g \tag{6.70}$$

V_T is the settling velocity of the sediment bed relative to fixed coordinates. $V_T C$ is the Kynch settling flux $\psi(C)$ extended into the region above the critical concentration C_c so that

$$F_d = \frac{(1-\varphi)V_r C}{\psi(C)} F_g \tag{6.71}$$

and

$$\frac{\partial \sigma}{\partial x} = -F_g\left(1 - \frac{(1-\phi)V_r C}{\psi(C)}\right) = -\Delta\rho g\varphi\left(1 - \frac{(1-\varphi)V_r C}{\psi(C)}\right) \tag{6.72}$$

The solid stress σ is assumed to be a function only of the concentration so that the rate of variation of the concentration with depth in the compressed bed is given by

$$\frac{\sigma'}{\rho_s}\frac{\partial C}{\partial x} = -\Delta\rho g\varphi\left(1 - \frac{(1-\varphi)V_r C}{\psi(C)}\right) \tag{6.73}$$

with $\sigma' = \partial\sigma/\partial\varphi$ the derivative of the function that defines the variation of the solid stress with composition.

The relative velocity can be related to the velocity at which the solids settles and to the total volumetric flux q.

$$q = \varphi V_s + (1 - \varphi)V_f \tag{6.74}$$

where V_s is the velocity of the solid and V_f the velocity of the fluid both relative to fixed coordinates

$$\begin{aligned} V_r &= V_s - V_f \\ &= \frac{1}{1-\varphi}(V_s - q) \end{aligned} \tag{6.75}$$

Substituting this into equation 6.73 and solving for the total solid flux

$$f(C) = V_s C = \frac{\sigma' \psi(C)}{\Delta \rho g C} \frac{\partial C}{\partial x} + \psi(C) + qC \tag{6.76}$$

Thus the dynamic differential equation that describes the behavior of the compressible sediment in a thickener is obtained from the differential mass balance, equation 6.2

$$\frac{\partial C}{\partial t} = \frac{\partial f(C)}{\partial x} = \frac{\partial}{\partial x}\left(\psi(C) + \frac{\sigma' \psi(C)}{\Delta \rho g C} \frac{\partial C}{\partial x} + qC\right) \tag{6.77}$$

Equation 6.77 is a non-linear parabolic partial differential equation that can be solved only numerically using initial and boundary conditions that are appropriate to the physical problem that is to be solved. Bürger *et al.* (1999) discuss solutions for a variety of important transient conditions in operating ideal thickeners.

Fundamental models for the effective solid stress in the compressible sediment have not yet been developed but the following empirical equations have been found to describe some typical compressible sediments.

Bürger *et al.* (1999) have used the following equation

$$\sigma(\varphi) = \begin{cases} \alpha e^{\beta\varphi} & \text{for } \varphi \geq \varphi_C \\ 0 & \text{for } \varphi < \varphi_C \end{cases} \tag{6.78}$$

where φ_C is the critical volume concentration and $\alpha = 5.35$ and $\beta = 17.9$.

Adorjan (1976) has used

$$\sigma = \begin{cases} a\left(\dfrac{\varphi - \varphi_C}{\varphi_U - \varphi}\right)^m & \text{for } \varphi_C < \varphi < \varphi_U \\ 0 & \text{for } 0 < \varphi \leq \varphi_C \end{cases} \tag{6.79}$$

where φ_C is the critical volume fraction and φ_U is the ultimate volume fraction of the sediment at which all compression stops. Adorjan (1976) has described an experimental apparatus that can be used for the direct measurement of the sediment compressive strength and has found values of $a = 182.6$ and $m = 1.81$.

6.7 Continuous thickening of compressible pulps

A particularly important solution of equation 6.77 is the steady condition that describes the behavior of operating industrial thickeners.

$$\frac{\partial}{\partial x}\left(\psi(C) + \frac{\sigma'\psi(C)}{\Delta\rho g C}\frac{\partial C}{\partial x} + qC\right) = 0 \tag{6.80}$$

which can be integrated to give the flux of solid at any level

$$\psi(C) + \frac{\sigma'\psi(C)}{\Delta\rho g C}\frac{\partial C}{\partial x} + qC = \text{constant} \tag{6.81}$$

The constant of integration can be evaluated by noting that at the bottom of the thickener the total flux must be equal to the convective outflow qC_D. Thus the concentration profile in the compression zone of the thickener is described by the differential equation

$$\frac{dC}{dx} = \frac{\Delta\rho g C}{\sigma'\psi(C)}(qC_D - qC - \psi(C)) \tag{6.82}$$

Hydrodynamic stability requires that $(dC/dx) \leq 0$ throughout the compression zone where $C_C \leq C \leq C_D$.

Equation 6.82 can be solved by numerical integration to give the steady-state concentration profile in the compression zone of the thickener. An example of this calculation is shown in Figure 6.14 which was generated

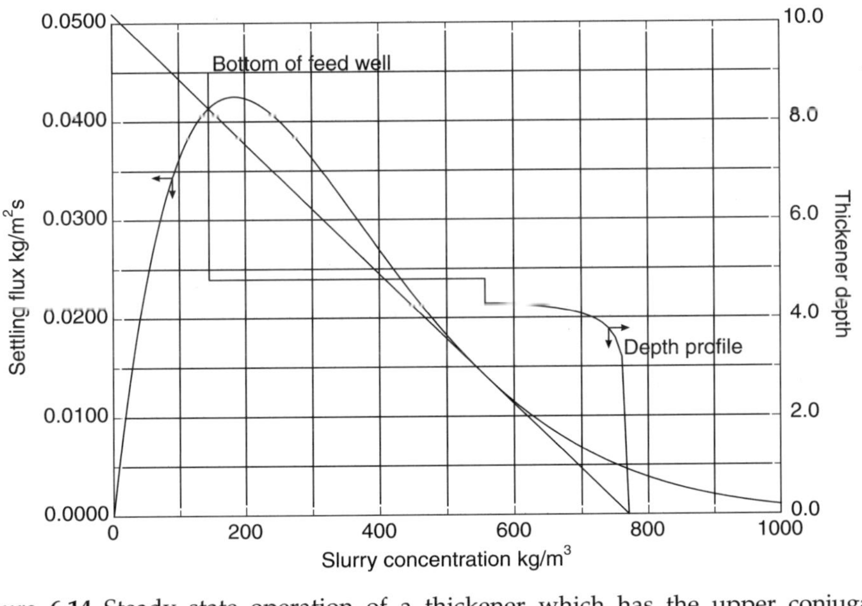

Figure 6.14 Steady state operation of a thickener which has the upper conjugate concentration lower than the critical concentration. The thickener conditions are defined in Illustrative example 6–1. The solid stress is given by equation 6.78 with $C_C = 575\,\text{kg/m}^3, \alpha = 5.35$ and $\beta = 17.9$

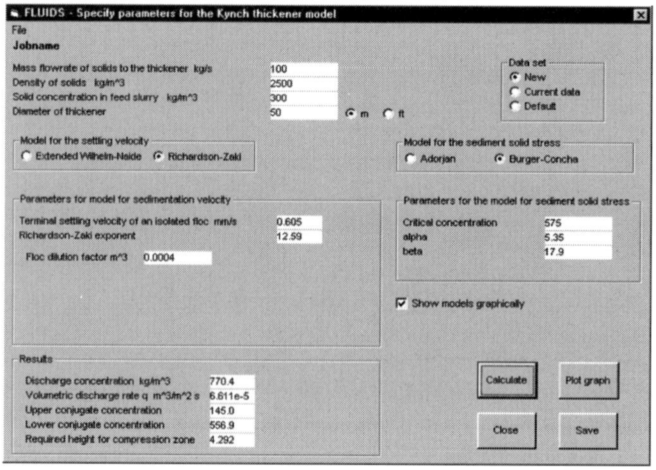

Figure 6.15 Fluids toolbox data input screen that generates figure 6.14

using the FLUIDS toolbox (see Figure 6.15). The reader is encouraged to reproduce this calculation.

Conversely the height of the compression zone that is required to achieve the required discharge concentration can be calculated from

$$\text{Height} = \int_{C_D}^{C_C} \frac{\sigma'\psi(C)}{\Delta\rho g C(qC_D - qC - \psi(C))} dC \tag{6.83}$$

The integral in equation 6.83 must be evaluated numerically which is not difficult to do although the integrand will have a singularity if $qC_D - qC - \psi(C) = 0$ within the range of integration. This means that an infinite sediment height would be required if this condition occurs. This in turn means that the lower conjugate concentration, C_M in Figure 6.10, which satisfies $qC_f - qC_M - \psi(C_M)$ cannot be situated between C_C and C_D. If that should occur, the discharge concentration must be decreased below the calculated value C_D by increasing the volumetric outflow q from the thickener discharge. A thickener that operates in this way has an operating line that falls entirely below the falling section of the flux curve as shown in Figure 6.16. The conditions that define the operation of the thickener that is illustrated in Figure 6.16 are (see Figure 6.17):

Mass flowrate of solids to the thickener = 60 kg/s.
Solid concentration in the feed slurry = 300 kg/m^3.
Density of solid = 2670 kg/m^3.
Diameter of thickener = 50 m.
Required discharge concentration = 1250 kg/m^3.

The extended Wilhelm-Naide model was used for the settling velocity with $V_T = 1$ mm/s, $\alpha_1 = 0.385$, $\beta_1 = 0.341$, $\alpha_2 = 6.2 \times 10^{-7}$ and $\beta_2 = 2.73$.

The Adorjan model, equation 6.79, was used for the solid stress with $C_C = 340$ kg/m^3, $C_U = 1296$ kg/m^3, $a = 182.6$ and $m = 1.61$.

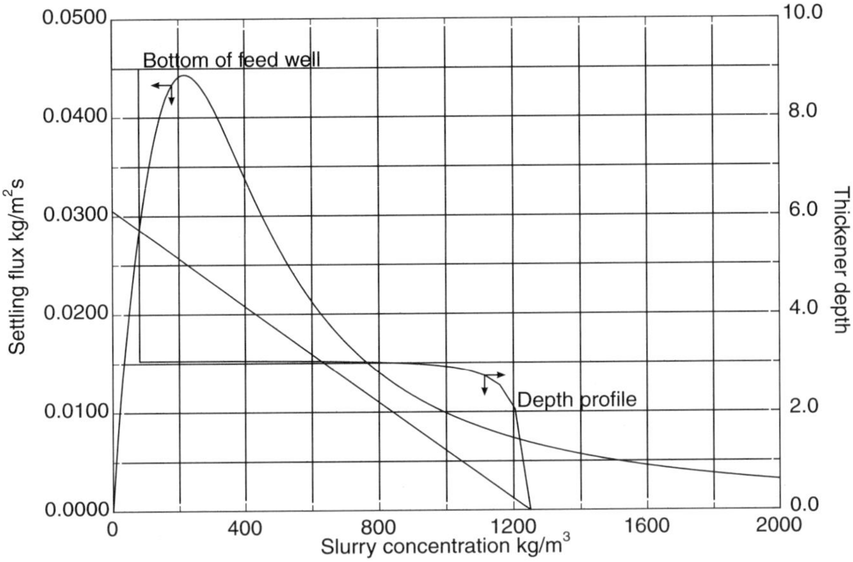

Figure 6.16 Steady state operation of a thickener which has the upper conjugate concentration greater than the critical concentration

The reader is encouraged to reproduce this operation in the FLUIDS toolbox and to investigate, for example, the way in which the height of the compression zone varies with the required discharge concentration.

If $C_C > C_M$ the desired discharge concentration can be achieved with the operating line tangential to the settling flux curve as shown in Figure 6.14. The vertical position of the discontinuity at the lower conjugate concentration C_M shown in Figure 6.14 is indeterminate and the discontinuity may not form at

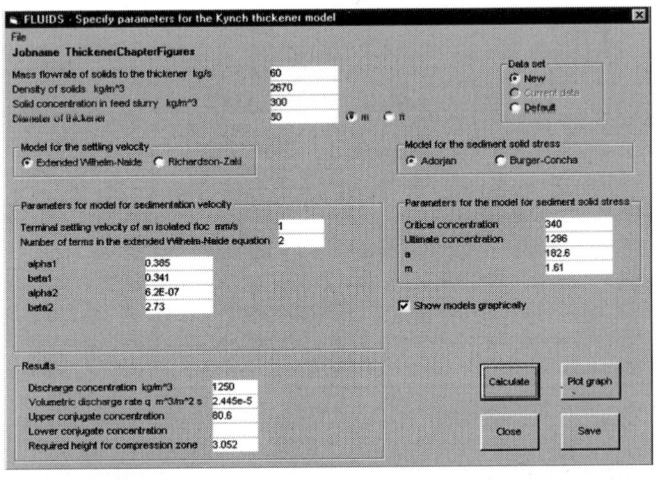

Figure 6.17 FLUIDS toolbox data input screen that generates Figure 6.16

all since it is only conditionally stable. If the discontinuity does not form there will be a sharp increase in concentration from the upper conjugate concentration to the critical concentration at 4.2 m from the floor of the thickener. The thickener operation that is illustrated in Figure 6.14 can be modified so that the operating line falls entirely below the flux curve. It is only necessary to increase the underflow withdrawal rate so that the discharge concentration decreases. The reader is encouraged to set this up in the FLUIDS toolbox and experiment with various settings of the underflow concentration.

6.8 Batch thickening of compressible pulps

The analysis of the behavior of a compressible sediment during batch settling is somewhat more complex than the behavior in the steady-state continuous thickener because batch settling is a dynamic process and the partial differential equation 6.77 must be used to describe the behavior rather than the ordinary differential equation 6.82. The solution to this equation is available in the FLUIDS toolbox. The user can choose between the Richardson-Zaki and extended Wilhelm-Naide models for the settling velocity and also between the Adorjan and Bürger-Concha models for the solid stress in the compression zone. A typical solution is shown in Figure 6.18 which was generated using the extended Wilhelm-Naide equation for the settling velocity and the Bürger-

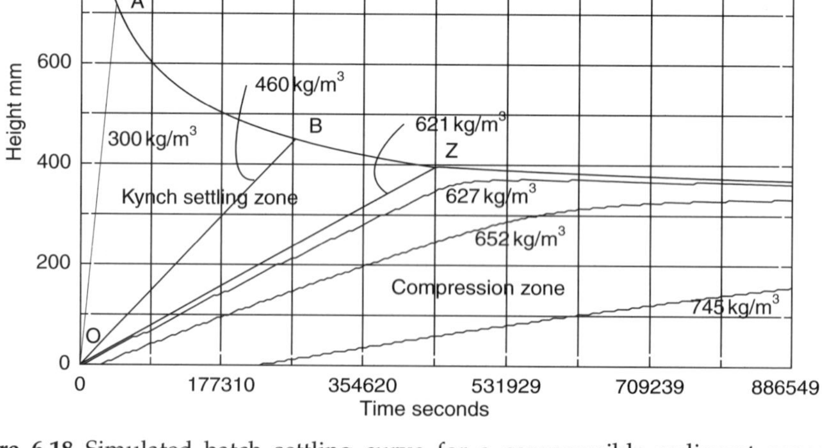

Figure 6.18 Simulated batch settling curve for a compressible sediment generated using the FLUIDS toolbox

Concha model for the solid stress. The parameters used for these models in the simulation are

1. Extended Wilhelm-Naide equation:

$$V_{\mathrm{T}} = 1\,\mathrm{mm/s},$$
$$\alpha_1 = 0.001,$$
$$\beta_1 = 1,$$
$$\alpha_2 = 10^{-10} \text{ and}$$
$$\beta_2 = 5$$

2. Bürger–Concha model:

$$C_{\mathrm{C}} = 621\,\mathrm{kg/m^3},$$
$$\alpha = 5.35\,\mathrm{Pa} \text{ and}$$
$$\beta = 17.9$$

The FLUIDS data specification screen for this simulation is shown in Figure 6.19.

The batch settling curve that is illustrated in Figure 6.18 has two zones: a Kynch settling zone above line OZ which is the contour of constant concentration equal to the critical constant C_{C} and a compression zone below this line in which the concentration is everywhere greater than the critical value C_{C}. Above line OZ the simple analysis presented in Section 6.4 can be used to calculate the settling behavior of the slurry. Line OA represents the propagation of a plane of concentration equal to the initial concentration C_0 in the settling vessel. This plane propagates upward from the floor of the vessel. The region OSA has uniform concentration C_0 and represents the initial constant rate period. The slope of the line SA is equal to the settling velocity of the solid at concentration

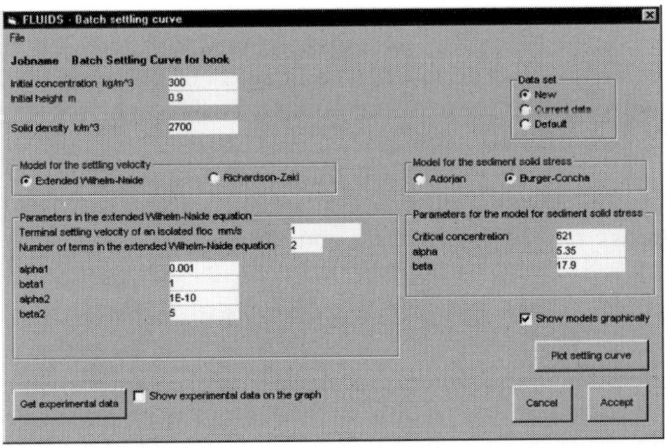

Figure 6.19 Data specification screen for simulating the batch settling of a compressible pulp using the FLUIDS toolbox

C_0. The region OAZ represents the falling rate period with the solid exhibiting Kynch behavior. Line OB is a typical concentration contour in this region and its slope is equal to the constant rate of propagation of the concentration plane. The region OAZ is filled with such linear concentration contours which are characteristics of the partial differential equation 6.2. The line OZ is the line of propagation of the plane of concentration equal to the critical concentration at which the flocs just begin to support each other by physical contact.

In the compression zone below line OZ the flocs are not settling freely because they are supported from below by sediment that has accumulated on the floor of the settling tank. The flocs are nevertheless settling because of the compression of the lower layers of sediment by the weight of the layers above. The rate of change of concentration at any horizontal level in the compression zone is given by equation 6.77. In the batch settler $q = 0$ so that

$$\frac{\partial C}{\partial t} = \frac{\partial \psi(C)}{\partial x} + \frac{\partial}{\partial x}\left(\frac{\sigma' \psi(C)}{\Delta \rho g C}\frac{\partial C}{\partial x}\right) \tag{6.84}$$

Equation 6.84 is a partial differential equation which describes how the concentration varies with time and position in the compression zone. In particular it describes how the concentration evolves with time from the corner of the region at the origin. Initial and boundary conditions are required to determine the solution.

The initial condition is $C = C_c$ at $x = 0$ and $t = 0$. On the upper boundary OZ, $C = C_c$ for $t > 0$ and $\partial C/\partial x = 0$ because the upper layer of sediment does not directly support any solid above by floc-floc contact.

On the lower boundary at $x = 0$, the flux $f(C, t) = 0$ since no solid passes through the floor of the container. Using equation 6.76 this implies

$$\frac{\partial C}{\partial x} = -\frac{\Delta \rho g C}{\sigma'} \tag{6.85}$$

The upper boundary of the compression zone moves with time. During the interval from $t = 0$ to $t = t_c$, the upper boundary increases at a constant velocity as described in section 6.5. The position of the boundary is therefore known at each value of t up to t_c and equation 6.84 can be solved numerically in a fairly straightforward manner.

At times greater than the critical time t_c, the upper boundary cannot be computed a priori since the pulp is settling by compression. The height of the sediment is governed by the integral condition

$$\int_0^{x_u(t)} C(x,t)\mathrm{d}x = C_0 h_0 \tag{6.86}$$

where $x_u(t)$ is the height of the upper surface of the compressible sediment.

Equation 6.86 reflects the requirement that all of the solid present in the vessel initially must be contained in the sediment after the critical time t_c.

Equation 6.84 is strongly non linear so that analytic solutions are not available and numerical solutions must be used. In practice an implicit finite difference scheme is used to ensure stability. The compression zone is covered by a grid so that at any grid point

$$x = j\Delta x \quad \text{with} \quad j = 0, 1, 2, \ldots, J = st \tag{6.87}$$

where s is the slope of line OZ and

$$t = i\Delta t \tag{6.88}$$

The grid spacings in the spatial and time dimensions are chosen to satisfy

$$\frac{\Delta x}{\Delta t} = s \tag{6.89}$$

so that grid points fall exactly and line OZ making the application of boundary condition on the upper surface convenient in the form

$$C_J^i - C_{J-1}^i = 0 \tag{6.90}$$

so that

$$C_J^i = C_{J-1}^i = C_c \tag{6.91}$$

The boundary condition at $x = 0$ is approximated by

$$\theta \frac{C_1^{i+1} - C_0^{i+1}}{\Delta x} + (1-\theta)\frac{C_1^i - C_0^i}{\Delta x} + \frac{\Delta \rho g C_0^i}{\sigma'(C_0^i)} = 0 \tag{6.92}$$

In the interior of the compression zone the parabolic partial differential equation 6.84 is approximated by the tridiagonal implicit scheme (Bürger and Concha 1998)

$$\begin{aligned} -\theta a_j^{i+1} C_{j+1}^{i+1} + \left(2\theta a_j^{i+1} + \lambda\right) C_j^{i+1} - \theta a_j^{i+1} C_{j-1}^{i+1} = {} & (1-\theta) a_j^i C_{j+1}^i - \left(2(1-\theta) a_j^i - \lambda\right) C_j^i \\ & + (1-\theta) a_j^i C_{j-1}^i + 0.5\left(\psi(C_{j+1}^i)\right. \\ & \left. -\psi(C_j^i)\right) \Delta x \end{aligned} \tag{6.93}$$

with

$$a(C) = \psi(C) \frac{\sigma'}{\Delta \rho g \phi} \tag{6.94}$$

and

$$\lambda = \frac{(\Delta x)^2}{\Delta t} = s\Delta x \tag{6.95}$$

A value of $\theta = 0.5$ appears to be satisfactory in most cases and the development of the concentration profile in the compression zone is easy to compute using equation 6.93 up to time t_c. For times greater than t_c, equation 6.93 is used but the height of the sediment decreases steadily so that equation 6.86 is satisfied at each time step.

When the concentration at each grid point has been calculated, the contours of constant solid concentration can be plotted by interpolating between grid points. Some typical results are shown in Figure 6.18.

The calculation methods that are presented in this chapter are not suitable for manual calculations and computer methods are essential. The FLUIDS toolbox provides a selection of useful models for both batch and continuous thickener calculations. The reader is encouraged to explore the different combination of models that are possible and to investigate the effect of parameter variations on the calculated results. It should be remembered that the models are highly nonlinear and the effect of even small changes in the parameter values can be quite dramatic. The influences of the various parameters are also highly correlated and interdependent. As a result not all combinations of the parameters will produce useful solutions.

6.9 Practice problems

1. The data shown in Table 6.1 were reported by a laboratory that specializes in dewatering technology. The data shows the interface height as a function of time for the standard batch settling test. The slurries were carefully

Table 6.1 Batch settling tests

20.0 per cent solids		*25 per cent solids*		*30 per cent solids*		*35 per cent solids*	
Time min	*Height mm*	*Time min*	*Height mm*	*Time min*	*Height mm*	*Time min*	*Height mm*
0.00	395.9	0.00	403.7	0.00	395.9	0.00	403.7
0.25	322.6	0.25	379.5	0.50	359.3	0.50	385.5
0.50	262.3	0.75	337.1	1.00	325.6	1.00	365.4
0.75	209.8	1.00	313.9	1.25	306.8	1.50	343.2
1.00	186.1	1.25	292.3	1.50	290.0	2.00	316.9
1.25	170.6	1.50	270.5	1.75	277.1	2.50	295.7
1.50	159.7	1.75	252.3	2.00	262.2	3.00	276.5
1.75	150.4	2.00	239.2	2.25	254.8	3.50	260.4
2.00	142.5	2.25	228.1	2.50	246.4	4.00	248.3
2.50	130.6	2.50	219.4	2.75	238.5	4.50	238.2
3.00	122.7	3.00	201.9	3.00	231.6	5.25	226.5
4.00	109.9	3.50	189.3	3.50	218.7	6.00	218.0
5.00	101.0	4.00	179.0	4.00	207.8	7.00	205.5
7.50	88.1	5.00	162.5	4.50	199.9	9.00	188.7
16.00	79.2	6.00	149.4	5.00	193.0	12.00	172.6
30.00	75.2	8.00	132.2	6.00	179.5	15.00	164.5
		10.00	121.5	8.00	160.3	20.00	157.0
		15.00	107.0	10.00	150.0	30.00	149.8
		20.00	102.9	15.00	132.0		
		30.00	100.5	30.00	118.8		

prepared by diluting the original concentrated slurry, adjusting the pH and adding flocculant at 0.02 lb/short ton calculated on a dry solids basis.

Use the Kynch construction to determine the settling velocity as a function of concentration. Do the data support the Kynch postulate that the settling velocity is a function of concentration only?

Fit the settling velocity results to one of the models that are described in the text and determine the maximum feed flux that can be handled by a thickener.

2. Use the toolbox to simulate the batch settling curves for each of the slurries whose settling velocities are modeled by the extended Wilhelm-Naide equations shown in Figures 6.7 and 6.8.

6.10 Symbols used in this chapter

A	Area. m^2
C	Solids concentration. kg/m^3 or per cent by volume.
C_C	Critical concentration. kg/m^3 or per cent by volume.
C_D	Discharge concentration. kg/m^3 or per cent by volume.
C_U	Ultimate concentration. kg/m^3 or per cent by volume.
C^+	Concentration immediately above a discontinuity.
C^-	Concentration immediately below a discontinuity.
F	Total flux in continuous thickener. kg/m^2 s or m^3solid/m^2 s.
h	Height of interface in batch settling test. m.
K	Permeability of floc bed. m^2.
n	Richardson-Zaki exponent.
r_F	Floc dilution. m^3/m^3
Q	Volumetric flowrate. m^3/s.
q	Volumetric flux. m^3/m^2 s.
t	Time. s.
V	Settling velocity. m/s.
v_{TF}	Terminal settling velocity of an isolated floc. m/s.
W	Solid flowrate. kg/s or m^3/s.
x	Vertical distance coordinate. m.
α	Model parameter.
β	Model parameter.
ρ	Density. kg/m^3.
φ	Volume fraction of solids. m^3/m^3.
σ	Velocity at which a discontinuity moves. m/s.
σ	Solid stress. Pa
ψ	Settling flux. kg/m^2 s or m^3solid/m^2 s.

Bibliography

The literature on sedimentation is large and is often contradictory. The treatment given in this book is based largely on the work of the joint University of Concepción – University of Stuttgart group which has provided a comprehensive analysis of sedimentation and thickening models. The publications of this group have been conveniently collected in a single volume (Bustos, *et al.* 1999) which is an invaluable source of careful mathematical analyses of many aspects of this fascinating topic. Numerical simulation of the dynamic behavior of continuous and batch thickeners is described in Bürger and Concha (1998) and Bürger *et al.* (1999).

Solid concentration profiles measured in operating industrial thickeners are presented in Stoltz and Scott (1972) and measured concentration profiles during batch settling of compressible sediments are given by Gaudin and Fuerstenau (1962).

References

Adorjan, L.A. (1976) Determination of thickener dimensions from sediment compression and permeability test results. *Trans. Instn. Mining Metall*, **85**, C157–C163.

Bürger, R., Bustos, M.C. and Concha, F. (1999). Settling velocities of particulate systems: 9. Phenomenalogical theory of sedimentation processes: numerical simulation of the transient behavior of flocculated suspensions in an ideal batch or continuous thickener. *International Journal of Mineral Processing*, **55**, 267–282.

Bürger, R. and Concha, F. (1998). Mathematical model and numerical simulation of the settling of flocculated suspensions. *International Journal of Multiphase Flow*, **24**, 1005–1023.

Bustos, M.C., Concha, F., Bürger, R. and Tory, E.M. (1999). *Sedimentation and Thickening: Phenomenological Foundation and Mathematical Theory*. Kluwer Academic Publishers.

Concha, F. and Bustos, M.C. (1991). Settling velocities of particulate systems, 6. Kynch sedimentation processes: batch settling. *International Journal of Mineral Processing*, **32**, 193–212.

Cross, H.E, (1963). A new approach to the design and operation of thickeners. *Jnl. South African Inst. Mining and Metallurgy*, **63**, 271–298.

Gaudin, A.M. and Fuerstenau, M.C. (1962). Experimental and Mathematical model of thickening. *Trans. Soc. Mining Engineers*, **223**, 122–129.

Ma, T.-W. (1987). Stability, rheology and flow in pipes, bends, fittings, valves and Venturi meters of concentrated non-Newtonian suspensions. PhD thesis, University of Illinois at Chicago.

Richardson, J.F. and Zaki, W.N. (1954). Sedimentation and Fluidization. *Trans. Instn. Chem. Engrs*, **32**, 35–53.

Scott, K.J. (1968a). Experimental study of continuous thickening of a flocculated silica slurry. *Industrial and Engineering Chemistry Fundamentals*, **7**, 582–595

Scott, K.J. (1968b). Thickening of calcium carbonate slurries, *Industrial and Engineering Chemistry Fundamentals*, **7**, 484–490.

Shirato, M., Kata, H., Kobayashi, K. and Sakazaki, H. (1970). Analysis of thick settling slurries due to consolidation *Jnl. of Chemical Engineering of Japan*, **3**, 98–104.

Stolz, E.C. and Scott, K.J. (1972). Design, operation and instrumentation of thickeners. Combined Report Chamber of Mines of South Africa. Project No. 11/504/64.

Turian, R.M., Ma, T.-W., Hsu, F.L.G. and Sung, D.J. (1997). Characterization, settling and rheology of concentrated fine particulte mineral slurries. *Powder Technology*, **93**, 219–233.

Wilhelm, J.H. and Naide, Y. (1981). Sizing and operating continuous thickeners, *Mining Engineering*, 1710–1718.

Index